Ways of Knowing in Science Series

RICHARD DUSCHL, SERIES EDITOR

Reforming Science Education:
Social Perspectives and Personal Reflections
Rodger W. Bybee

STS Education: International Perspectives on Reform
Joan Solomon and Glen Aikenhead, Editors

STS Education

International Perspectives on Reform

Edited by
Joan Solomon
Glen Aikenhead

TEACHERS COLLEGE PRESS
Teachers College
Columbia University
New York and London

Published by Teachers College Press, 1234 Amsterdam Avenue
New York, NY 10027

Figure 5.2 is reprinted from an article appearing in Volume 31, Number 1 of the journal *Theory into Practice,* entitled "The Integration of STS into Science Education," by G. S. Aikenhead (1992). Reprinted by permission of the publisher (*Theory into Practice*).

Figure 14.3 is reprinted from an article appearing in Volume 17, Number 3 of the *British Educational Research Journal,* entitled "Teaching about the Nature of Science through History: Action Research in the Classroom," by J. Solomon and K. Harrison (1991). Reprinted by permission of the publisher, Carfax Publishing Company, Abingdon, United Kingdom.

Library of Congress Cataloging-in-Publication Data

STS education : international perspectives on reform / edited by Joan Solomon, Glen Aikenhead.
p. cm. — (Ways of knowing in science series)
Based on papers presented at an international STS symposium, held in 1990 at Oxford University.
Includes bibliographical references (p.) and index.
ISBN 0-8077-3366-0 (acid-free paper). — ISBN 0-8077-3365-2 (pbk.: acid-free paper)
1. Science—Study and teaching. I. Solomon, Joan. II. Aikenhead, Glen S. III. Title: Science-technology-society education. IV. Series.
Q181.S79 1994
507.1—dc20 94-10822

ISBN 0-8077-3366-0
ISBN 0-8077-3365-2 (pbk.)

Printed on acid-free paper

Manufactured in the United States of America

01 00 99 98 97 96 95 94 8 7 6 5 4 3 2 1

Contents

Preface

This book celebrates the fact that science-technology-society (STS) education in school science has come of age, particularly in Europe, North America, and Australia. A richness of STS journals, organizations, research projects, curricula, textbooks, and conferences attest to the newly achieved status of the STS movement. One of us first suggested producing a book about STS education as long ago as 1984, but it was too soon. Then, in 1990 at an international STS symposium at Oxford University (Developing a Relevant Science Education for Future Citizens), we revived the idea, encouraged by the scholarly standard and interesting diversity of the contributions at the symposium. We realized the time was now right for a book on STS education for school science (STS science education). This book, although by no means a collection of those symposium papers, grew out of the Oxford meeting. STS science education was ready for its first substantial book of scholarly essays from the international community.

Writing about STS education has its complexities. As John Ziman acknowledges in Chapter 3, the subject is confusing because "the fundamental purposes of STS education are genuinely and properly diverse and incoherent." Similarly, David Layton in Chapter 4 writes that STS "has never been a precisely defined innovation, and it has been driven by an intriguing mixture of motives." This mixture of motives and intentions, perhaps appearing incoherent at times, has been driven by strong social forces over the past four centuries and particularly over the past 50 years. Part I of this book describes these social forces from several different perspectives, showing how times have changed dramatically and as a consequence, how STS science education is eager to enter new educational territory.

Throughout a short and turbulent history, the academic field of "social studies of science" (STS studies) has also been driven forward by social forces. Jean-Jacques Salomon, one of the first scholars in this field, wrote in 1973 that the study of STS was "born of war." That was how it looked in Europe. In the same decade, however, American philosophers such as Theodore Roszak (1972) were arguing that science studies had emerged from "the rape of the environment." The educational movement based on STS has always been radical and critical of old assumptions. Both STS studies and STS education roll forward on the waves that sweep through society, whether they are those of war or of the devastating environmental changes produced in times of peace. Although there has been far too little interaction between the STS studies and education communities, both work for changes in the perception of science and technology to match the actual roles of science and technology in society.

Within the STS education community itself, two features unite the field but also break it wide open: its relevance to contemporary society and its acceptance that the status quo must change. To urge that science education should match the times in which our young people live is certainly to set foot on an STS journey, but there are several pathways to take. The young in one country will be driven by different motives from those in other countries because the aims and problems of their societies are different. Whereas traditional science education can be monolithic, and may set out to teach the orthodox concepts and laws of science from pole to pole, the STS educator's commitment to local and contemporary relevance precludes a monolithic approach.

It is easy to see the difficulty for some scientists and science educators to accept STS education with its regional mix of motives and intentions. If there is one social feature of science on which scientists have always prided themselves it is the norm of universality. Scholars in the various disciplines of science have always traveled from university to university secure in the knowledge that their concerns and their ways of thinking and working, even the equipment that they use, would fit perfectly with those of their colleagues on the new campus. "Science itself," they are fond of saying, "is an invisible college." Does STS education mean an end to the perception of science as universal?

These issues are particularly hard for anyone attempting to compose and edit a volume of essays about STS education for school science, such as this one, which spans so many different countries. And yet if there is a deep purpose to STS education, we can only hope to discover its nature by examining STS education against different backgrounds. Every science educator seems sure that STS education has its own determining direction or directions, although few have been able to articulate what those may be. In an attempt to find an answer we need to look to the field of science studies, and by contrasting this field with STS education, find uniting purposes for STS education in school science.

Under the umbrella of "science studies" there are a number of well articulated sub-disciplines such as the history and philosophy of science and technology, and also their sociology, economics, and politics. The latter group (sociology, economics, and politics of science and technology) is referred to as STS studies. The overall purpose of STS studies has always been to tackle, in an intellectual sense, the problems of science and technology in a social setting. Variations in these settings only add grist to this academic mill. STS studies are all concerned with human communities, so that by changing the conditions under which the communities live and work, by moving the focus of study from one country to the next or from one epoch to another, these situations offer valuable opportunities for testing out theoretical descriptions under different circumstances. STS studies might also help in tackling, or monitoring ways of tackling, inherent social and economic problems.

Within the broad field of science studies, many of the research questions that scholars pose for themselves are internally generated from the nature of science. They are about the scientific process, or how the community operates its publication system and deals with anomalous results. Contemporary issues from the outside world are rarely the major focus of a field of study, they only become examples to illustrate a theory. In the words of Spiegel-Rösing (1977):

> Consequently what they try to solve in most cases are cognitive problems (to know how science functions) rather than operational problems (to know how science can be governed or directed). (p. 17)

The focus may have moved a little more toward the operational mode in the last two decades, but science studies are still essentially inward looking (Pinch, 1992). Hence the united purpose of tackling the problems of science and technology *in a social setting* is little more than a blurred outline. Nevertheless, the academic community of science studies has certainly provided scholarship. Indeed the prominent STS educators who contributed to this book have, almost without exception, quoted from the works of the science studies community. This adds an intellectual richness to their educational scholarship. However, as long as the science studies community looks only inward it cannot contribute very much to STS education.

The substantial difference between the science studies community and that of STS science education concerns action. The severe problems emerging from the impact of technology on our ways of living are now analyzed and reported at length in academic journals, for example, *Science, Technology, & Human Values* or *The Social Studies of Science*. Prestigious conferences such as those sponsored by the "4S" (Social Studies of Science and Society) movement are committed to intellectual programs that now include more external issues such as medical ethics and public resistance to nuclear power. Nevertheless, the science studies community rarely takes direct responsibility for public action (Pinch, 1992). That responsibility is passed on to economists, politicians, and scientists. On the other hand, STS science education presents a field for direct professional action that is open to innovation. The *Bulletin of Science, Technology & Society* attempts to bridge the gap between the two communities. Only a small number of academics in science studies have contributed to STS science education through collaborative efforts with educators. John Ziman and Gerald Holton are examples, as are some of the science studies scholars who actively participate in the National Association for Science Technology and Society, whose Technological Literacy Conference is held annually in Washington, D.C. (described by Bill Williams in Chapter 6).

The purposes of STS education are broader in scope than those of the academic science studies community. The focus of both is still the problems of science and technology within society. However, the aims and motives of the educationalist lie partly in the inside world of science, partly in the outside world of society, and partly in the world of students. The day-to-day intention of STS teachers is to teach science, and teach it well. According to them this can only be done in a meaningful way if it is carried out in a social or technological context. Teachers speak of motivating students and of preparing them to meet the science-based social problems that abound outside the world of school.

In their usual pragmatic fashion, science teachers are less interested in academic science studies than in how students should best be taught. The problems of science and technology in a world of human communities add a great deal of color, human interest, and student involvement to science lessons. Once teachers see this in the eyes of their students, the battle for change is won. Teachers know the worth of the STS approach when they know how well it teaches science to their students.

Somehow, any volume about STS education must convey the pragmatic approach to STS as well as its scholarship. These two aspects, pragmatism and scholarship, are at the heart of the purposes of STS science education.

This book has little of the purely academic approach found in science studies. Yet almost every chapter shows some reflection on the historical, philosophical, or sociological aspects of science and technology. Part I, "Rationales and Challenges," is perhaps richer in science studies content than most of the other parts, but even here the call to action is sounded loudly by Glen Aikenhead in Chapter 2. Part II contains five chapters dealing with "STS in Science Classrooms." This is at the pragmatic heart of the action, yet the scholarly backing from the field of science studies is evident.

The content of Part III, "Public Understanding and Cultural Diversity," strongly reflects the changing nature of STS. Some years ago, the science studies community was arguing about the magical beliefs of the Ashanti and whether or not those beliefs constituted a science. Today the STS education community knows that teachers must consider local values and public understanding in both Western and developing countries. It is no longer an academic question. Young people in all countries want to understand their environment and to care for it by action. The values that they put on knowledge to achieve such understanding make an enormous difference to how they approach learning science.

Part IV has two chapters devoted to "Gender Questions and STS." To understand the social context of science and science education, the gender anomalies and enigmas must be studied both inside and outside the classroom. Here is a topic that we judged to stand on the boundary of science education and science studies. Chapter 15 by Joan Solomon provides research-based insights from inside the classroom, whereas the outstanding feminist piece by Hilary Rose (Chapter 16) captures a radically different view of science from outside the classroom. Hilary Rose's chapter lies in the science studies domain more than the other chapters in the book.

The book's final part, "Research into STS Education," brings out the pragmatic nature of our claims to knowledge about STS education. Here there are no tales of individuals' personal constructions of science and technology, nor are there new developmental theories about learning. In their place are carefully described innovations where learning, teaching, and researching go hand-in-hand. A review and critical discussion of STS research round out Part V.

Our book illustrates how the context and content of STS science education may vary from a leaf infestation in India, to industrial effluents in the United States, to radiation contamination over Europe, to expert testimony in a Canadian courtroom, to science and technology in Australian industries. The clear intention is to deal with problems of the interactions of science, technology, and society. The expert STS educationalists who wrote for us needed no cues to avoid the pitfalls of academicism or political extremism. Throughout the book their concern for young people's science education supersedes every other aim.

Our only disappointment arises from the limitation of the book's length, and consequently from not being able to include contributions from other prominent educators who would have brought us an even more encompassing view of STS education from more countries and from more facets of science education.

Each part of the book begins with a short description of its chapters. Although each part addresses only one theme, the reader is invited to read more deeply into each chapter to discover a wealth of coherent messages about what STS education in school science means.

Joan Solomon
Oxford University

Glen Aikenhead
University of
Saskatchewan

PART I

RATIONALES AND CHALLENGES

STS has had to fight hard for its place in science education and must continue to do so today. Although the debate over STS education varies from one country to another, the main arguments in those debates are reflected in the four chapters in Part I.

The first two chapters come from opposite sides of the Atlantic. Each begins with the history of European science to exhibit the deep roots of STS, but each continues along a separate track. Joan Solomon (Chapter 1) doubts if there has ever been a "valid," "neutral," or "contemplative" science since the time of Francis Bacon. Certainly the effect of two world wars and the environmental shocks of the 1960s revealed the falsity of any such notions. As a consequence, she argues, current social issues and the discussion of social justice must be admitted into the science classroom via STS instruction. Glen Aikenhead (Chapter 2) describes how the facade of respectable neutrality was built to protect the image of the institution of science. This image was particularly strong in the 19th century when the high school science curriculum was established. Social events of the 20th century changed science's image but not the antiquated 19th century school science curriculum. STS science teaching corresponds to 20th century science.

John Ziman's special contribution to Part I (Chapter 3) is his recognition of the educational diversity that STS has been called upon to tackle. He shows that no one purpose can have pride of place. It is the teacher, rather than the content, that characterizes STS science teaching.

Part I ends with three challenges to the STS movement by David Layton (Chapter 4): to determine the place of technology in STS curriculum, to explore the implications for citizen participation and empowerment, and to tackle the problem of cultural diversity. Peter Fensham and Deborah Corrigan (Chapter 18), for instance, address the issue of technology, whereas Joan Solomon (Chapter 10) and Anita Rampal (Chapter 13), among others, write about citizen action. Part III of this book is devoted entirely to cultural diversity.

In the domain of STS education there are always new challenges and new questions. This book was written in an attempt to provide some of the answers.

Chapter 1

Conflict Between Mainstream Science and STS in Science Education

Joan Solomon

This chapter advances the argument that the present conflict between the educational traditionalists and the advocates of STS has grown out of a much deeper and more ancient controversy that has been latent within science itself for more than three centuries. The argument is essential to proper understanding of science education on both sides of the Atlantic. When the discussion turns to the eventual development of STS education, the emergence of STS is traced in the context of British educational policy.

THE SCIENCE FOR THE STATE

When the foundations of British science were being laid in the 17th century, there was a vigorous attempt to make science an instrument of the state instead of a branch of metaphysics or Greek philosophy The originator of this movement was Francis Bacon, Lord Chancellor of England and founding spirit of the Royal Society. In his book *The Advancement of Learning,* Bacon (1605) wrote with scorn about the kind of academic education he had received, common at that time, as "no more than a conducted tour through the portrait gallery of the Ancients" (p. xi). Bacon dedicated this book (perhaps the earliest ever to be written on the virtues of a top-down kind of STS education) to King James I, and was astute enough to recommend science as being useful. The preface and much of the text is too fulsome for a modern taste, but there is no mistaking the argument that Bacon is making. He scoffs at the idea that learned men (scientists) should live a life of remote contemplation unsullied by concern for civic matters, by which he means the wealth and welfare of the nation.

> But the greatest error of all the rest is the mistaking or misplacing of the farthest end of knowledge . . . [which is] sincerely to give a true account of their gift of reason *to the benefit and gift of men*. This is that which will indeed dignify and exult knowledge, if *contemplation* and *action* be more nearly and straightly conjoined and united together [emphasis added]. (p. 35)

There was nothing new in recommending the applications of science to the rulers of states, but in Bacon's program for science the perspective was broad. He was claiming that there was an unfilled place waiting for the scientific enterprise and for education's role in the enterprise within official policy and action on behalf of the state.

Some time later when Thomas Spratt, historian of the Royal Society, wrote to James II on the same subject he brought up the Plague and the Great Fire of London as the sort of public disasters that action based upon science might have been able to avert. Although new at that time, this kind of agonized response to immediate disaster, or topical problems, is very familiar to us all today. It calls for scientific action, and the call comes from lay people, not from scientists.

Bacon had been shrewd enough to argue for both contemplation and action, or as we might put it now, for pure science and for applied science. The balance between these two was to characterize the swings from so-called "valid" science to "relevant" STS, and visa versa, in the centuries that followed, right up to the science education curricula of the present age.

TOP-DOWN SCIENCE

Despite Bacon's agenda, mainstream British science has traditionally resisted both politics and practical technology. For years it looked only to the Royal Society for leadership. One might claim that this has proved to be both too abstract and too elitist for the healthy growth of science. Robert Hooke, for example, was from lower class origins and spent much of his early scientific career making air pumps and other mechanical devices for more upper crust scientists like Robert Boyle to use in demonstrations to the Royal Society.

In the 18th century the inspired engineers of steam power had little education in science. Some of the innovative engineers, such as the Cornishman Trevithic, could not even sign their names to patent forms. The abstract theory of thermodynamics arose out of the mathematics of a French savant, Sadi Carnot, who considered the operation of an idealized engine more than a century after the real steam engines had been set to work. The practical inventiveness of America's Thomas Edison or Oliver Evans was not missing in 19th century Britain, but their British counterparts rarely found their way into the higher circles of the scientific establishment.

As long as science remained a semiaristocratic top-down pursuit, which it mostly did during Newton's time and for the two centuries following, it did not need to consider the cause of the common citizen. Science was a reflection of the contemporary culture, just as the community of scientists today reflects our own more democratic contemporary culture.

If it had been the system of science that had divorced the contemplation of pure knowledge from active technology, then it had clearly been responsible for holding up the progress of science itself. It had divided the making of instruments and machinery from the thinking about them, just as it divided the population into those who might advance scientific thought and those who would "only" make technical devices. Most of all science ignored those members of society who simply used the prevalent technology.

Top-down science also held back other kinds of progress. The Industrial Revolution produced pollution on a scale we can hardly imagine today, and there was no one to use scientific knowledge for controlling it. Early prints of Ironbridge, where coal was first used in the steel-making blast furnaces, show a hideous landscape half hidden by yellow sulphurous fumes the acidity of which can only be guessed. By the middle of the 19th century the life expectancy in Manchester was a mere 28 years. The new chemical industry made alkalis such as sodium and potassium hydroxide by profitable new methods, and in the process, huge quantities of hydrochloric acid gas were released, making an acid rain of pH 2 (as compared to the pH 4 of today's acid rain). There was almost no comment on social issues from the learned and well-heeled scientists, most of whom had not contributed to the new mechanical and chemical progress. Consequently, there was no recourse for those unlearned in science who had to operate the new machines or live in the areas devastated by the acid rain or strong acid.

By the end of the 19th century there would be universal education, but even then there was precious little science in what was being taught. The question that the early 20th century had to address was whether the public—society as a whole—had any role to play in science, and whether science as system of thinking was sufficient in itself to give the ordinary citizen, who had no intention of becoming an academic scientist, something of value that should be included in their education.

NEW NEEDS FOR CITIZEN SCIENCE

In the post–World War II years, before the new STS courses began in schools, two very substantial areas of need emerged for "citizen science" in education.

One of these was public awakening to the environmental effects of new technologies. While books like Rachel Carson's *Silent Spring* (1962) drew attention to the quiet slaughter of birds and butterflies and horrified many in the United States, Britain lost much of her ancient heritage of hedges and wild flowers. Moreover, the introduction of a virus to kill the wild rabbit population by means of myxomatosis resulted in horrific scenes in the countryside where blind and dying rabbits crawled around in slow death-throes. Chemical and nuclear wastes produced pollution that most scientists and industrialists chose to ignore. It was important to note that there was also little or no legislation to provide the ordinary citizen with access to information about these science- and technology-based hazards.

But information was clearly not enough. People needed an understanding of science if they were to express an opinion on the quality of their environment. This was the first call for a science education that would be appropriate for the ordinary citizen.

A different incentive for a new kind of science education came indirectly from an influential report by a group of the world's top intellectuals, economists, and business people in The Club of Rome. This body had commissioned a report about the economics of future global developments that would produce "more understanding of the world's predicament" and "stimulate new attitudes and policies" to address them. The report, *The Limits to Growth* (Meadows, Meadows, Randers, & Behrens, 1972), started a debate that included items such as exponential growth in fuel use, the finite nature of the fossil fuel reserve, the world population explosion, and the limited production of food.

Shortly afterwards, the energy crisis of 1973–1974 raised the price of petroleum so that the cost at the gasoline pump quadrupled in a single year. These events stimulated a series of pessimistic television programs of a type that became known as the "doomsday syndrome." This was noteworthy because the new type of public or citizen science was to use the medium of television as its "text" far more than any schoolbook (see my discussion of this in Chapter 10).

The scientists were still keen to bring their knowledge to bear on world problems. In the 1960s, a group of American plant breeders working in Mexico had developed strains of new "miracle grains"—wheat and rice—that launched the Green Revolution. This was designed to enable the Third World to feed itself and seemed at first to be a remarkable success.

Some scientists were optimistic about solving other problems too. They focused attention on the new technologies of alternative and nuclear power. They placed the problems of the developing world on the agenda of science. If science and its technology were to solve these problems, as some seemed to believe, then this might be a new area of activity for science. At the very least it provided an attractive new way of teaching the traditional concepts of science through their applications.

TOP-DOWN *STS* EDUCATION

At first, the only references to the difficult concepts of the global "Limits to Growth" debate in schools were to be found in questions on advanced level General Studies papers, or on Oxford and Cambridge scholarship entrance examinations. This was a part of the school that was specially privileged, so it produced an intellectual debate that was not seen as having much to do with everyday technology in the developed world. It is not surprising, therefore, to find that the first scheme in Britain to be entirely devoted to STS—*Science in Society* (Lewis, 1981)—was generated in the exclusive public schools for the sixth-form age group (age 17 to 19 years). Problem solving in a Third World context soon became a popular activity in science lessons at some British schools, perhaps because of the high moral implications of "doing good" for the unfortunate in less developed countries.

But there were other characteristics of this early form of STS that served to separate it from later STS courses. It was based on the premise that scientific thinking of the traditional kind could solve all the social problems in the STS arena. These teachers had been involved with the introduction of the new syllabuses of the 1960s and 1970s. (See Andrew Hunt's account of this in Chapter 7.) They were enthusiasts for scientific thinking in a way that made them very suspicious of environmental movements or any other "bottom-up" pressure groups that did not base their case on respect for traditional "valid" science. British science textbooks had always included a short final paragraph in each section on "the applications of science." This practice served to increase respect for science and its knowledge, which was thereby shown to be not only certain but also useful and beneficial. The proponents of the earliest STS courses in Britain taught that science was about "facts" and that its history showed a trail of unmitigated progress and triumphs.

This attitude toward the problems of society was dubbed "scientism" by John Ziman (1980, and in this volume, Chapter 3). In educational terms, scientism tended to trivialize STS, but fitted in well with the high academic status of the schools involved in the educational initiative. This was not citizen science but top-down remedies from the scientific elite (many of whom had contributed to booklets in the series *Science in Society*) on behalf of the less fortunate. In many ways this STS approach remained firmly in the tradition of establishment science, even though the texts now included such topical issues as renewable energy resources, industrial management and nuclear power. A similar circumstance in the United States with *ChemCom: Chemistry in the Community* is discussed in Chapter 9.

STS AND EDUCATIONAL RHETORIC

In general education, there have always been a variety of aims for students and different perspectives on the purposes and problems of schooling. Somewhere among these aims and perspectives, the emerging STS movement would have to claim a place. By the 1970s this was not difficult to do. Young (1971), for example, had examined the educational debate about equality of opportunity, and had moved its focus from "selection by ability" to "hidden messages" contained in the curriculum itself. In science education it was particularly easy to see the force of this analysis: Both the traditional and the newer Nuffield courses in the United Kingdom of the 1960s were about "valid" science for the intellectually able.

Large-scale educational surveys dating from this period (for example, Peaker, 1967) had shown that the major influence on pupil achievement in school was not so much the material conditions of the home (although children from the lower socioeconomic groups certainly exhibited lower achievement levels than those from rich homes) as the educational aspirations of the parents. It seemed to follow that high achievement in science would only come about if the general population wanted and valued it.

Some years later Skilbeck (1984) wrote about the general purposes of education for "social reconstruction" as well as for "cultural transmission." Although his argument was not specific to science it certainly fitted the scientific case. Cultural transmission corresponds to traditional contemplative science education, whereas social reconstruction was a substantially new goal for science. Its terminology implied that every generation has a right to act, within the constraints of the democratic process, in order to reshape its own society. If an absence of scientific knowledge prevented ordinary people from thinking and acting on issues about which they cared deeply, then science education had seriously failed. The criteria for success and failure were no longer simply factual knowledge or examination grades. School science education, according to the new agenda of STS, had a duty to empower the action of future citizens.

Both of these educational arguments (equality of opportunity within the curriculum and citizen empowerment) were tantamount to an invitation to curriculum reform. STS, with its emphasis on social responsibility, was needed inside the curriculum to complement the more traditional approach to science education.

A NEW BREED OF SCIENCE EDUCATION COURSES

The first mainstream school science course anywhere to take up the challenge to include a strong STS component was SCISP (Schools Council Integrated Science Project, 1973). It satisfied several of the strands in STS by teaching directly about economics and industry, as well as about scientific method. It achieved only a small following in the schools, but arguably quite a large influence elsewhere.

Educational progress rarely comes about in a logical progression, especially not in the old-style British education system that had been (at least up until the Education Reform Act of 1988) one of the most flexible systems in all the world. It was lightly driven by an examination system applied to students only at age 16. By the 1960s, it had even become possible for school teachers to construct their own syllabus and examination—the Certificate of Secondary Education. In 1981, one of the first STS courses, *SISCON-in-Schools* (Solomon, 1983), chose this simple way of breaking into schools and their examination structure.

Earlier in 1978, *SISCON-in-Schools* was begun by the few schoolteachers who had attended a summer school course about the university version of the SISCON project (the precursor to *SISCON-in-Schools*). Like the earlier *Science in Society* (Lewis, 1981), this newer scheme was for sixth-formers, but it differed in the strands of STS it chose to take up (Solomon, 1981). Where the earlier course was strong on economics and industry, *SISCON-in-Schools* focused more on the human and fallible nature of scientific theories, the quality of life in the First and Third Worlds, and the social effects of new technologies, including armaments. It was also intended for, and taught to, the full ability range of students.

SOCIAL JUSTICE AND MORAL RESPONSIBILITY

There was another sense in which the new STS teaching was very different from what had gone before. Traditional contemplative science had not only removed itself from the worlds of action and citizenship, it had also claimed the moral high ground of abstract thought. Indeed, some of the old advocates of science in the curriculum had claimed that the rigors of logical thought and the discipline of experimental verdict could produce the best training for a good moral life. It was not dissimilar to the scientism described by John Ziman in Chapter 3.

All belief in the moral neutrality of science should have vaporized in the searing heat of the Hiroshima explosion. This was by no means the first time that scientific research had been directed at the production of weapons, but now it was no longer possible to excuse the scientists from blame on the grounds that others had misused their inventions. Top scientists from all over Europe had flocked to Los Alamos to work, quite knowingly, on what they called "our bomb." When it was all over, many realized that it had not been just duty or patriotism that had made them work so hard at constructing this new and devastating weapon, but the basic scientific challenge itself. Robert Oppenheimer, who led the team making the atomic bomb, once said that whenever a problem looked technically "sweet," there would always be scientists ready to work on it. Oppenheimer's statement suggests that the scientific enterprise itself might be to blame. Science and scientists have responsibilities towards society. In STS courses those responsibilities would be encountered in the classroom.

This is another point of conflict between the traditionalist contemplative science educators and the new breed of STS educators. The former see the science classroom as a haven from the untidy unjust world of emotions, and they use the classroom to teach logical truths. But the danger is that the public, including school children, translate that lofty picture of scientific inquiry into a cartoon image where amoral scientists pursue their intellectual goals at the expense of animal and human welfare. The traditionalists argue that moral education is all right in its proper place, in the humanities classroom, but not in science. The STS teachers reply that citizens and future scientists alike need to marry the intellectual study of science to feelings for moral right and social justice. Science and feelings cannot just be stitched together after the uncaring cartoon image has already been planted in the minds of children.

In the contemplative tradition of valid science education, as of science itself, there is neither a sense of personal duty nor of social responsibility. But where relevance and action are central to the aims of science and science education, neither of these aspects of morality can be avoided. Perhaps STS teachers actually welcome their inclusion, seeing their commitment to children's learning painted in the broadest brush strokes, where the ability to consider the moral outcomes of their actions, whether scientific or not, is in the forefront of all educational endeavor.

Chapter 2

The Social Contract of Science: Implications for Teaching Science

Glen Aikenhead

In his 1939 book, *The Saber-Tooth Curriculum,* Harold Benjamin sketched a parable of stagnate curriculum reform. Let me quote a synopsis:

> Three fundamentals marked the first educational curriculum: (1) catching fish with bare hands, (2) clubbing tiny horses to death, and (3) frightening saber-toothed tigers with torches.
>
> By studying those three subjects in their "schools" the stone-age people got along fairly well until there came a changed condition caused by the movement of ice from the north, the forerunner of the ice age.
>
> The streams became muddied and fish could not be seen to catch with the bare hands, so someone invented the net, made of vines. The tiny horses fled and the antelope replaced them. The stone-agers invented antelope snares. The saber-toothed tigers died of pneumonia, but the big ice bear replaced them, and the stone-age men dug pits to trap them. So net-making, twisting antelope snares and digging bear pits became the three essentials of life.
>
> But the schools continued to teach fish-catching with the hands, horse-clubbing, and tiger-scaring because they had taught them for years. Some "liberal" wanted to teach net-making, snare-making, and pit-digging but he was met with opposition. Some even wanted to do away entirely with the old subjects, but they aroused a storm and were called radicals.
>
> The old subjects must be retained for their "cultural value," the school people contended. The proposed new subjects had no place in the curriculum.
>
> The conservatives said: "Training to catch non-existent fish with bare hands is the best way to achieve muscular coordination and agility; training in clubbing horses that do not exist is an education in stealth and ingenuity; practicing to frighten tigers that do not exist develops courage. Some things are fundamental and sacred in education and must not be changed." (Benjamin, 1948, pp. 53–54)

This chapter develops the argument that the high school science curriculum, as normally taught today, is a saber-tooth curriculum. The curriculum was established in the 19th century, and although times have changed dramatically, the fundamental and sacred aspects of the 19th century science curriculum remain with us today.

The argument proceeds in two interconnected parts. The first part traces the evolution of science from its status as "natural philosophy" in Renaissance society to its modern status in the 1990s. The evolution of science demonstrates how social events have shaped the very nature of science itself. This historical analysis establishes a framework for the second part of the argument, which demonstrates how social events over the last three decades will likely reshape the high school science curriculum. The argument concludes that an STS science curriculum will replace the saber-tooth curriculum, for the same reason that over 100 years ago science replaced natural philosophy.

The historical analysis explaining the evolution of science draws heavily upon a series of lectures given by Everett Mendelsohn (1975a, 1975b, 1975c, 1976), professor of history of science at Harvard University. This historical analysis of science provides new insights into today's high school science curriculum. The analysis gives a context for understanding why science teachers tend to teach as they do, why educators are calling for a change in this teaching, and why an STS curriculum is the natural focus for such a change.

THE EVOLUTION OF SCIENCE

Over the past 500 years, historical events have helped to shape the very nature of science itself (Cohen, 1960; Dampier, 1948; Elkana & Mendelsohn, 1981; Mendelsohn, 1975a, 1975b, 1975c, 1976; Middleton, 1963). Three instances will be examined:

1. How the social context of 17th century Europe, dominated by the Counter-Reformation, gave birth to the *institutionalization* of natural philosophy (the "science" at that time);
2. How the social context of 19th century Europe, dominated by the Industrial Revolution, precipitated the *professionalization* of science (when the term "science" came into common use);
3. How the social context of the 20th century, dominated by World War II, molded the *socialization* of science (the "science" of our everyday world today).

These three episodes caused mutations to science's fundamental characteristics. These critical episodes will be examined, one at a time, to see how they guided the evolution of science and how they profoundly altered science's social contract with society.

Our story begins with the Renaissance, a time when Aristotle's philosophy of the world governed people's conceptions of matter, motion, and the heavens. Some scholars added incrementally to Aristotle's intellectual heritage; for example, Bernard Sylvester (12th century), Roger Bacon (13th century), Nicole Oresme (14th century), Leonardo da Vinci (15th century), Copernicus (16th century), and finally, Johannes Kepler and Galileo Galilei (early 17th century). Their individual efforts began to define a new type of philosophy—*natural* philosophy. Natural philosophy rejected authority based on scripture and old philosophers, in favor of authority based on one's empirical experience with nature.

The Renaissance explorers of nature (da Vinci, Kepler, Copernicus, et al.) were unimpeded by any social contract with their society. They were only hobbyists enjoying intellectual diversion. Their musings were not seen initially as interfering with traditional philosophy. As the 17th century approached, however, the political climate changed. The Reformation led to the Counter-Reformation. A social contract would have to be negotiated between natural philosophy and society.

The Institutionalization of "Science"

During the first half of the 17th century, a number of natural philosophers worked towards organizing themselves into a politically acceptable public enterprise. Leaders in this effort included Mersenne, Descartes, Bacon, Huygens, Boyle, and Hooke. In 1662, British natural philosophers were formally incorporated into The Royal Society. They were followed in 1666 by the French natural philosophers who formed the Academie de Science in Paris. What content did they include in their new type of knowledge? What content did they leave out? What were the social forces at the time? How did these forces determine what would be included, and what would be excluded?

First, let us examine the social forces. Seventeenth century Europe was a time of insecurity and anxiety. People were coping with the Counter-Reformation, wars, fires, and epidemics. It was a period of instability—social, intellectual, and political instability. Traditional sources of authority had been undermined. Natural philosophers of earlier generations had contributed to this revolution. Some had been condemned. Some had even been burned.

In 1660, however, a new social order was in place. Cromwell's rule had ended in England. The church and crown were reestablished. World exploration and colonialization had given rise to a new mercantile middle class.

What did natural philosophers offer this new society in return for allowing the natural philosophers to institutionalize their new way of thinking? What compromises did the natural philosophers have to make?

The promise that natural philosophy offered society was best summarized by Francis Bacon when he wrote about three kinds of ambition. The third and most noble ambition was for man "to establish and extend the power and dominion of the human race over the universe. . . . We cannot command nature except by obeying her and understanding her" (quoted in Mendelsohn, 1975c, p. 9). In short, knowledge is power! "Scientia est potentia." Not only is knowledge power, but within the domain of Christendom, with its Judeo-Christian ethic, you could exploit nature with a mood of indifference to the feelings of natural objects. Therefore, natural philosophers offered 17th century society three gifts: (1) a new way of knowing characterized by a new type of authority, an authority based on observation and rationalism, and not on scriptures and social position; (2) the goal and ability to achieve power and dominion over nature; and (3) a mood of indifference toward any responsibility to nature or to those who might be affected by the new rational knowledge.

But there was a compromise. In order to establish a niche in society, natural philosophers had to make peace with the newly established religious and secular authorities. The compromise was clear. Natural philosophers would avoid discussing religion, politics, and morals. These topics would be excluded, along with subjectivity and arational thinking. The public face of The Royal Society was therefore established. The social contract of natural philosophy was finalized: natural philosophy would deal only with objective rational knowledge acquired through direct experience with nature, and in return, natural philosophy would provide other social institutions with power and dominion over nature. Mendelsohn (1975b) calls it the positivist compromise.

Natural philosophy, as a social institution, was established across Europe. Its social contract with 17th century society kept natural philosophy out of trouble with the new authorities. Moreover, natural philosophy provided a new middle class with useful, practical knowledge that carried no moral responsibility. To be sure, discussions about moral responsibility were defined by the social contract to be beyond the scope of natural philosophy. In short, the precursor to science had been declared objective and value free, because its political survival depended on it.

The social and historical events of the 17th century helped to shape the characteristics and limitations of what we call science today. In the meantime, other historical events would reshape the nature of science.

The Professionalization of "Science"

The 17th and 18th centuries saw natural philosophers gain power and dominion over nature. By the end of the 18th century, their successful techniques and knowledge were redirected by others—technologists—towards power and dominion over *human productivity* itself. This gave rise to the Industrial Revolution and gave new power to the social institution of technology. Technology became so successful that it challenged the social niche that natural philosophy had gained 200 years earlier. Industrialists saw natural philosophy as the handmaiden of technology. Natural philosophers would have none of it.

They reacted to the attempted subordination in several ways: by retreating into the cloisters of the universities, by calling themselves "scientists" (to distinguish themselves from natural philosophers), by creating a public face of "pure science," by isolating themselves from the "vulgarities of practical knowledge," and by establishing a tight rein over who would have access to becoming a scientist and what standards would apply. Natural philosophy had evolved into a profession.

The Industrial Revolution caused natural philosophy to redefine its boundaries and to renegotiate its mission in society. By redefining its boundaries, what did science include? What did it exclude?

Science focused its efforts on intellectual curiosity and knowledge for knowledge's sake, a marked departure from its Baconian tradition of practical knowledge. Moreover, science distanced itself further from value-laden discourse, from the consumers of its knowledge, and from social responsibility. Science eschewed its technological and social connections.

Scientists established a self-serving hierarchical position by defining technology as "applied science." This misconception continues to plague technology and science education today (Collingridge, 1989; Fleming, 1989; McGinn, 1991; Snow, 1987).

By 1860, a reshaped domain of knowledge had been constituted. Biology, chemistry, geology, and physics were enshrined as disciplines when they became new administrative units within the university. Natural philosophy had become professionalized science.

Coincidentally, the high school science curriculum was being introduced into public schools for the first time. The university's administrative model for science was copied by the high schools. Biology, chemistry, geology, and physics became the only valid ways to view nature. Like 19th century science, high school science eschewed practical knowledge and ignored values and social relevance.

Similar to science itself, the high school science curriculum was shaped by the social forces that existed at the time of its inception. This happened to the high school curriculum when science was retreating into the universities to protect itself from a take-over bid by technology. As a consequence, the legitimacy of school science was defined by the 19th century professionalization of university science, and the purpose of school science was to prepare students for university science.

The Socialization of Science

The 20th century brought a host of new social forces. World War II likely reshaped science more than any other single historical event. World War II ensured the marriage of aloof scientific expertise with the life-or-death practical problems of technology. This unlikely marriage irrevocably bound science and technology into a strong social unit called research and development (R & D). This marriage necessitated a new social contract between science and society.

The transformation of science during World War II is epitomized by one of the most dramatic events that occurred: the production and deployment of the atomic bomb. It was a corner around which humanity turned. Science constructed the corner and guided humanity around it. The isolation that scientists by and large had enjoyed since the 19th century was evaporated by a mushroom-shaped cloud.

By the end of World War II, "small science" had become "big science" (Price, 1963). Big science had profound implications. It meant big budgets; large partnerships with government, industry, and the military; and a narrowed gap between "pure" and "applied" science. Big science meant the creation of national wealth and military superiority. As a result, scientific knowledge today has political currency on two levels: (1) internationally, where it is traded in the diplomatic halls of foreign policy (Dickson, 1984); and (2) nationally, where it sustains the dominant socioeconomic intrastructure of that society (McGinn, 1991). For instance, governments support R & D in order to maintain their country's competitive edge in the world marketplace (Ziman, 1984).

Government, industry, and the military have become the dominant patrons of scientific activity. Science of the 1990s occurs in an interactive world of politics, economics, and war. Only a small minority of academic scientists undertake pure research. Even these scientists are mindful, however, of the political lobbying required to obtain funds. One conclusion seems inescapable: the social significance of scientific knowledge now takes on a new 20th century reality.

While World War II was having an impact on the interactions between science and society, a new academic discipline emerged—the sociology of science (Layton, Chapter 4; Ziman, 1984). Anthropological studies into the social construction of scientific knowledge (for example, Latour &

Woolgar, 1979) described two types of science: "public science" and "private science" (Holton, 1978). Public science is the science reported in journals, at conferences, and in textbooks. Private science, on the other hand, is what actually occurs in labs. It is recorded in personal notebooks, conversations, E-mail, and letters. What did the anthropologists discover about scientists? In contrast to the public face of science—that objective, rational, open-minded, free communication, and honest face, which we recall was established in response to the social demands of a 17th century Europe—private science was found to harbor subjectivity, arational thought, closed-mindedness, secrecy, and behavior less than honest (Gauld, 1982).

Today we recognize two social contexts of science; an *external* context in which science interacts with technology, economics, politics, law, ethics, and other facets of society; and an *internal* context in which historical and social dynamics mediate the production of knowledge (Rosenthal, 1989; Ziman, 1984).

Science still strives for power and dominion over nature, but in a new context of research and development where technology, values, and social responsibility play an increasingly important role (Mendelsohn, 1976). Thus, a new social contract between science and society seeks a balance between, on the one hand, power and dominion over nature, including economic well-being, and on the other hand, stewardship of the earth and quality of life.

Summary

Science today differs dramatically from the "science" of 1660 and 1860. Each stage in the evolution of science has been shaped and reshaped by social forces, both outside and inside science. This is schematically represented in Figure 2.1 by the arrows between "social forces" and "science." The social forces of 1600 gave birth to the institutionalization of science (box I). The social forces of 1800 precipitated the professionalization of science (box P). And lastly, the social forces of the 20th century molded the socialization of science (box S).

Figure 2.1 serves as a framework for the next section. The argument turns to the social forces of the last three decades, and the consequence of those social forces in reshaping the high school science curriculum.

THE HIGH SCHOOL SCIENCE CURRICULUM

As described above, the social context of 17th century Europe shaped the institutionalization of science. Similarly, the social context of the 19th

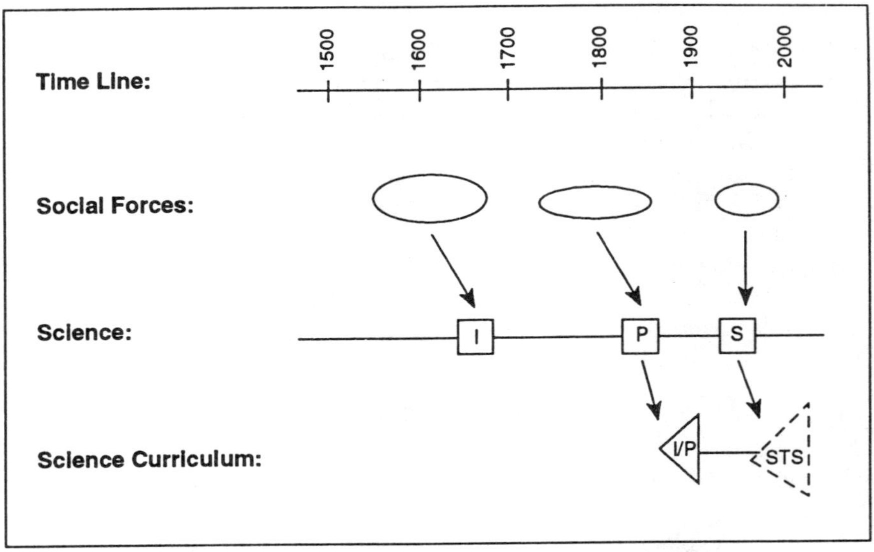

FIGURE 2.1 *The influence of social forces on science and the science curriculum*

century shaped the fundamental tenets of the high school science curriculum. Figure 2.1 depicts this influence by the arrow from box P (the professionalization of science) to triangle I/P (the institutionalization and professionalization of the high school science curriculum). Figure 2.1 provides a framework with which we can understand current practices and anticipate future changes.

What can we understand about the science curriculum given its 19th century origins? We can recognize why today's curriculum includes pure abstractions that demonstrate the aesthetic unity of the disciplines, and why practical knowledge and social concerns are all but excluded. We can see why the curriculum portrays science as a purely rational and objective inquiry into absolute knowledge. We can now recognize this portrayal as the public facade of 19th century science. We can also appreciate how the social upheavals of the 17th and 19th centuries shaped this facade.

Moreover, we can now recognize that this facade of school science—19th century academic idealism—is inconsistent with the realities of post World War II science. In other words, school science is seriously out of date. Like the saber-tooth curriculum, the school science curriculum embraces outmoded content and values. No wonder we hear the criticism that school science is sterile, false, and boring (Science Council of Canada, 1984).

Several serious attempts have been made to modify the high school curriculum in North America over the past 50 years (Hurd, 1986). Educators have tried to replace the curriculum's 19th century academic isolationism with a 20th century authenticity that reflects the humanization and socialization that science itself has undergone. But every attempt has failed (Hurd, 1986). The "conservatives" (as Benjamin called them) with vested interests in the saber-tooth curriculum possessed greater political power than the "liberals" who tried to implement an up-to-date curriculum.

Nevertheless, we have reason to be optimistic about the latest struggle to modify the science curriculum. During the past three decades a number of new social forces have become evident. Synergetically they may cause the extinction of the saber-tooth curriculum.

John Ziman explains in Chapter 3 that student motivation in the sciences has decreased, as evidenced by a depletion in enrollment and an apparent attrition in academic achievement. Research tends to show that school science actually *discourages* imaginative and creative students, particularly women and minorities, from entering the profession (Bondi, 1985; Majumdar, Rosenfeld, Rubba, Miller, & Schmalz, 1991; Oxford University, 1989). School science, therefore, undermines the development of future scientists and engineers. This is evident in the United States with its nation-at-risk crisis, in which the decline in both enrollment and student capabilities threatens to compromise that nation's place in the competitive market of our global village (Hurd, 1989; Majumdar et al., 1991). Industry and labor support science and technology education in order to maintain a sound economy (Bondi, 1985; De Vore, 1992).

The environmental movement has raised the public's consciousness concerning the stewardship of the earth. In ubiquitous conflicts between corporate profits and environmental agendas, scientists have been seen to participate on all sides of an issue (*Globe & Mail,* 1983; Jacobson, 1983). As a consequence, the attentive public has changed its perception from an image of science in a context of objective isolation to one of science with social agendas.

Furthermore, two new academic fields have articulated the social nature of science. The *social studies of science* field explores the internal sociology of science, whereas the *science policy studies* field concentrates on social issues external but related to science (Spiegel-Rösing, 1977). For instance, David Layton in Chapter 4 discusses the social studies of science and how the "Social Responsibility in Science" movement activated STS programs at universities and colleges. In Chapter 5, I show how the two fields help define STS teaching.

Concomitantly, a "science-for-all" education movement has surfaced in many countries (Fensham, 1992). The science-for-all impetus in Canada, for instance, took the unusual form of a national science education policy calling for a socially relevant science curriculum (Science Council of Canada, 1984). A science-for-all education contrasts with the elitist view of education of science for the few. Joan Solomon traced the historical roots of this contrast in Chapter 1. "The few" refers to students who prepare for, and survive, the university screening mechanism that artificially selects students who look very similar to the very academic scientists who sustain that screening mechanism (Tobias, 1990).

Practical capability (Harrison, 1980; Layton, 1991) is the most recent social issue to challenge science education, as described by David Layton in Chapter 4. Today's pressure to synthesize science and technology education is not unlike the earlier 20th century pressure that merged science and technology into the institution we call research and development (R & D).

In summary, the past three decades have witnessed the emergence of substantial social forces: (1) a pervasive decline in the interest and understanding of science; (2) an awakening recognition of science as a human, social, and technological endeavor; (3) an egalitarian movement in public education; and (4) a proposal to synthesize science and technology education. Each country has its own unique set of social forces that impinge upon its 19th century school science curriculum. Many countries are beginning to change their traditional science curriculum.

When designing a new curriculum, countries share a common trend toward teaching science embedded in technological and social contexts familiar to students (Bybee, 1985a; Eijkelhof & Kortland, 1988; Fensham, 1992; Hofstein, Aikenhead, & Riquarts, 1988; National Science Teachers Association [NSTA], 1982; Piel, 1981; Ziman, 1980). This new curriculum movement advocates teaching science in a science-technology-society (STS) approach. An STS science curriculum conveys to students an image of science that honestly reflects science's social character—the 20th century socialization of science.

This new curriculum is depicted in Figure 2.1 by the STS triangle. The socialization of science (box S) constitutes a pervasive pressure on today's school science curriculum. The arrow between box S and triangle STS represents this pressure. The triangle has broken lines, indicating a tentative status. Will the social forces of the past three decades, including the social nature of science itself, establish a "socialized" science curriculum—an STS science curriculum? How much longer can 19th century school science masquerade as legitimate science?

Reform has already begun, as evidenced by the successful initiatives discussed throughout this book. STS science educators are closing the gap between 19th century school science and 20th century authentic science.

Chapter 3

The Rationale of STS Education is in the Approach

John Ziman

A bit more than 10 years ago, I felt so confused about STS education that I decided to write a book about it. I did not expect that this would be of much use to other people, but at least it would help me to get my own ideas straight. In the event, the first expectation was realized but not the second. *Teaching and Learning about Science and Society* (Ziman, 1980) has sold just enough copies to satisfy my publishers, but there is no sign that it has had any effect on STS education, and my own thinking on the subject is still very unclear. A lot has happened in the 1980s, but still not enough to bring the theme of this chapter into focus in my own mind.

I hope that you will not feel cheated by this confession of perplexity. Perhaps that is the first lesson to be learnt on this whole subject: the movement for STS education springs from so many different sources, and flows in so many different channels, that it does not have a shape that can be grasped mentally and described as a whole. That is not necessarily a defect. The same would apply to other great movements of our times, such as those for peace and for the environment. Such movements are kept alive by countless personal and collective commitments that cannot be captured in a few thousand words of didactic prose.

NOT NEGATING SCIENCE BUT EXTENDING IT

Of course, the simplest way of characterizing STS education is to contrast it favorably with what I have called "valid" science education; that is, science as it is conventionally taught in schools and universities. This a relatively easy exercise to which I devoted several chapters of *Teaching and Learning about Science and Society.* The target is large, lumbering, and very much the same the world over. After all, valid science is supposed to be universal, so surely it should be taught the same way everywhere.

The objection to this disparaging attitude is that it imparts a negative, destructive, almost antiscientific taint to the STS enterprise. Such a taint is entirely inappropriate and undesirable. It is not just a matter of keeping in the good books of the scientific mandarins, who are sometimes overzealous in defending the good name of their calling. It is much more a matter of making it quite clear that STS education genuinely acknowledges the manifold accomplishments of science, and seeks only to enhance, complement, and extend them. To put it another way: *the fundamental weakness of "valid" science as it is usually taught is not what it says about the world, but what it leaves unsaid.* The task of STS education is to fill that gap.

A MULTIPLICITY OF APPROACHES

We can now see why there are so many different approaches to STS education. This is not a sign of deep divisions and sectarian conflict within the STS movement. The different approaches are not contradictory but complementary. Each approach is designed to bridge the gap, so to speak, along some particular sector of the moat surrounding "valid" science. It tries to introduce the student to some particular aspect of science in its social context, and thus to complement and extend conventional science education in that particular direction.

STS education has not developed holistically. Some educators have felt the need to develop a particular curriculum and to teach it in a particular way because they have not been satisfied with a specific aspect of conventional science education. Each approach thus embodies a particular line of argument for STS education and is worthy of serious consideration as such.

This is not to say that there is necessarily a coherent whole, an overarching rationale, of which these different lines of argument are interlocking segments. On the contrary, having started from quite unconnected points of view, it would be a miracle if they did not sometimes lead to contradictory conclusions. Even if there were a scholarly consensus on the nature of science and technology in themselves (which there certainly is not) there would be profound disagreements about the nature of the society or societies in which they are embedded. The fact that STS education is bound to convey messages about society itself introduces a controversial political element, which is unsettling for conventional teachers of "valid" science, but which brings the whole subject to life for receptive students.

The fundamental purposes of STS education are genuinely and properly diverse and incoherent. Nevertheless, it is instructive to consider each of the principal approaches to STS education in turn, indicating its supposed objectives and drawing attention to some of its main advantages and disadvantages.

For my own convenience, I have gone through these in the same order as in Chapter 7 of *Teaching and Learning about Science and Society*, but somewhat more schematically. Indeed, I am not suggesting that the arguments I am presenting are particularly novel, or that this particular survey covers the whole subject. What we need at this stage in our discussions is a checklist, a catalogue raisonné, a rough conceptual framework, within which we can locate the rich variety of experience and opinions reported in later chapters of this book.

The Approach Through Relevance

The most natural way of extending "valid" science is towards its technological and other practical applications. Indeed, our notion of the solidity, the reliability, the ultimate validity of much scientific knowledge, rests on its successful worldly use. The principal justification for the support of basic research by society is its potential utility. To draw attention to the relevance of science, in detail, to everyday life, is to provide it with a clear social role.

Pedagogically, this is a relatively open and easy path to follow. There is a long tradition of illustrating scientific concepts by their applications. Sometimes these are rather stereotyped or artificial; for example, in the famous problem requiring a mathematical analysis of the sliding of a spherical elephant down a weightless ladder leaning against a perfectly smooth wall. But there are innumerable possible starting points, ranging from everyday matters of weather and cooking to the sophisticated enchantments of chemical industry and space transport.

There can be little objection to the proposition that people ought to know something of the science behind the familiar aspects of life, and something of the aspects of life that science has affected. From a broader STS perspective, however, this approach is inadequate. It takes a quick stride from science to technology, but then it usually fails to go on to society. It is not enough to demonstrate that scientific knowledge is not self-sufficient, an end in itself, but is relevant to some other domain of life.

It is necessary also to place that other domain (that is, science-based technology) into a wider context. Indeed, this approach, by failing to draw attention to the impotence or irrelevance of science in dealing with certain aspects of life, may foster a *technocratic* attitude, where it is assumed that society should be shaped and ruled solely according to scientific or technological principles.

The Vocational Approach

What might be called the vocational approach to STS education comes more naturally at the university than at the school level. Many students in institutions of higher education are studying scientific or technological subjects as a preparation for careers in scientific research, engineering, medicine, or other science-based callings. It is obviously desirable that their education should prepare them for their future social roles, as well as instructing them in purely technical matters.

The rationale of such an approach is clear, and there can be no doubt of its value to those for whom it is designed. It has always been one of the weaknesses of professional education to concentrate on its theoretical aspects at the expense of the realities of practice. Many of the skills that need to be acquired are organizational in a narrow managerial sense, but they do make the student aware of social and ethical dilemmas that need to be rehearsed before they are encountered on the job.

Although courses such as these are often quite unconnected with other initiatives for STS education, they penetrate quite a way into the society element, and raise critical questions about the social role of science and technology. In many respects, these questions are about the social roles of scientists and technologists—as advisors, expert authorities, repositories of knowledge, or responsible decision makers. A study of these roles, with particular attention to their ethical implications, can thus be a point of entry to a more thorough exploration of the whole subject.

Although few students at the secondary level already envisage themselves in these roles, the fact that the issues at stake can be presented in personal terms makes them accessible to a much wider group. Thus, for example, a TV drama enacting the everyday dilemmas of a biomedical researcher or a vet could be just as instructive as a whole series of lectures on the ethics of animal experimentation.

The objection to the vocational approach, as normally practiced, is that it tends to take a narrowly professional line, emphasizing the positive role of the doctor, the engineer, the researcher, or other technical experts, with little reference to the downside that might be perceived by other people. This is probably unavoidable because one of the main functions of vocational education is to develop a commitment to professional standards, both technical and social. A specialist has to believe quite strongly in the value of what he or she is doing in order to do it well.

Society needs its specialists, and cannot afford to disillusion them at the same time as it is training them to do their jobs. This is not to suggest that the purpose of STS education is to expose illusions supposedly engendered by professional scientific and technological education, but it is not very helpful to take a line that brings the student very close to this very tempting target without the intention of subjecting it to searching criticism.

The Transdisciplinary Approach

One of the notorious characteristics of science is that it is highly specialized. Even at quite an elementary educational level it is customarily divided into distinct disciplines. One of the major movements for reform in science education in recent years has been to break down these barriers. Although transdisciplinarity as such is not usually regarded as one of the major routes into STS education, it does have close links with it.

What the transdisciplinary approach emphasizes is the overall unity of the natural sciences and their associated technologies. All too often a student is taught, in quite unrelated lessons, to think about, say, a piece of stone solely in terms of physics on Monday, solely in terms of chemistry on Wednesday, and solely in terms of geology on Friday. What the student should be taught is to see this as a natural object, all of whose properties are interrelated, even if it is convenient to explore them in different ways. The goal of integrating the sciences, and presenting scientific knowledge as a unity, or at least as a seamless web, is no longer metaphysical or impractical. It is clearly justified from a philosophical point of view, and is fully in keeping with the way that science has progressed in the 20th century.

This holistic conception of science is surely an essential, basic principle for STS education. It must be taken as given, for example, in any discussion of the extent to which practical problems arising in society can be tackled scientifically. The social aspects of an advanced technology cannot be fully appreciated without an understanding of the transdisciplinary mixture of scientific knowledge that has gone into its making. If the social sciences are to be brought in for the study of the society leg of the STS tripod, then they must be shown to be connected with the natural science and part of the same system of thought.

Transdisciplinarity as a central pedagogic principle thus provides a possible approach to STS education. It presents a correct account of the nature of scientific knowledge in terms of its actual content. It tells us to sketch out a picture, however crude and schematic, of what we think we know scientifically about the world.

But this is an incomplete rationale, with limited scope for STS education. It does not have much to say directly about technology, and only looks at society as a distant object of study, not as the place where we are at. And if one is not very careful one may present such a strong and definite image of unified science that it may seem impervious to uncertainty or criticism. Once more, a tinge of "scientism" permeates this whole approach.

The Historical Approach

Although scientific knowledge is supposed to be superior to the accidents of history, research is a profoundly historical activity. Research is dedicated to progress. It strives to change what we learned in the past into something new and different. One of the central norms of science is originality. Of their very essence, science and technology are dedicated to an historical process of creating novel modes of thinking and living.

The history of science is not, therefore, just a fascinating story of human enterprise. It is an indispensable dimension for any understanding of the nature of science itself. STS education must encompass this dimension. It must show that science and technology grow and change in association with the societies in which they are embedded. It must show the increasingly influential role of science and technology in society, and the increasing demands of society on science and technology. The historical approach, by showing how things got the way they are, is one of the most compelling ways of explaining this present state and also of laying out the ground for a discussion of how it might change.

From a more narrowly pedagogic point of view, the history of science and technology provides excellent case studies of processes still at work in contemporary society. Famous episodes such as the trial of Galileo, the public debate on *The Origin of Species,* or the tale told by James Watson in *The Double Helix* exemplify in clear and simple terms important themes in the social relations of science. These themes are difficult to dissect out of present day complexities. Of course, the presentation of such an episode is bound to be greatly oversimplified and distorted by the enchantments of historical distance and hindsight, but that need not nullify the elementary STS lesson it was chosen to illustrate.

For reasons such as these, the history of science has long been regarded by scientists and other academics as the most natural medium for humanizing science education. Unfortunately, this enthusiasm has not always been very fruitful. From the point of view of STS education at the secondary school level, there are two major disadvantages to this approach.

The first disadvantage is that the actual history of science is peculiarly deep, subtle, and complicated. Every famous episode is really only a nodal point in an immensely complex web of people, ideas, items of evidence, cultural influences, political and economic forces, and institutional forms—all unfamiliar and changing from generation to generation.

Those of us who have had the nerve to enter this endless forest are often fascinated by it, and spend more time than we can spare in exploring corners of it in detail. But the average student, especially the would-be scientist or engineer, is not attracted to such studies in the first place, and is very easily repelled by this intellectual complexity. In a word, the history of science is just too academic, too remote from the everyday world, too rarified emotionally, and sometimes far too pedantic, to catch the imagination of most young people and draw them into STS education.

The other disadvantage is that at an elementary level the history of science is very Whiggish, very bullish, about science. To make any sense of what is going on at all, the student must first have a basic framework of the principal discoveries and their discoverers, in chronological order. It is almost impossible to present such a chronicle except as a celebration of scientific progress. Even when the account includes outmoded concepts, such as phlogiston or caloric, the final triumph of truth over error can be guaranteed.

There is also an inevitable tendency to concentrate on the achievements of a few individuals, attributing to them heroic or saintly qualities far above normal humanity. This hagiographic attitude is, or course, the dominant ideology in the scientific community, which is obsessed with the notion of individual creativity, even when the only evidence of this is the luck of being the first to report a significant discovery. Such an attitude is quite inconsistent with an STS education designed to present science as a collective enterprise, a social institution, which is not fueled solely by the psychological traits of a few exceptional or fortunate people.

Philosophical Approaches

One of the principal objectives of STS education must be to give an account of the *nature* of science. Unfortunately, this is a much more controversial matter than many scientists appear to believe. But the traditional view is that this account should be essentially philosophical. Science should be presented as a more or less coherent body of knowledge, organized logically around theoretical principles and validated by observation and experiment.

The great virtue of this approach is that it does attempt to tackle head on the question that scientists themselves carefully skirt around: is science true, and if so why, or if not why not? In the end, STS education has to explore the grounds for belief in science, because this is the ultimate basis of its social influence, in technology and beyond.

Unfortunately, as we all know, the philosophy of science is a very difficult subject of byzantine complexity and unplumbed depth. It is true that even secondary school students can be brought to an understanding of many elementary points, such as the general characteristics of scientific explanation, the formulation of theories and their testing by experiment, and the provisional nature of all scientific knowledge. Unpedantic presentation and informal discussion of such matters should surely be a feature of all science education from the very beginning.

But as soon as this approach is made more formal it runs head on into the antipathy of the great majority of students to abstract concepts and arguments. A genuine taste for philosophy is rare, even among mature scientists. They have no patience with the quibbles and fine distinctions that are required to make a subtle point, and do not feel the necessity of digging deeper and deeper in search of a firm base on which to build.

In any case, it is not clear what STS education has to gain from the philosophy of science as currently expounded in academic circles. Beyond a certain point, it seems to fragment into a variety of esoteric and conflicting schools of thought, whose teachings do not seem to cohere into a consistent scheme. The principal lesson from the current stalemate on the philosophical front is that this is not a fruitful area for detailed study under the STS banner!

More seriously, from an STS perspective, the philosophical route does not lead anywhere near society. It is directed towards the development of a purely intellectual conception of science, where thinking somehow takes place outside a real world of thinkers, actors, and talkers. In my opinion, this "mentalism" is misconceived. I cannot understand how it could ever be possible to develop a meaningful epistemology of science (a general theory of the nature and status of scientific knowledge) without a *social* dimension. Science is a highly institutionalized *communal* activity, not the internal monologue of a lonely seeker after truth.

The Sociological Approach

The very notion of teaching about science, technology, and society clearly suggests a sociological approach. It calls directly for teaching about science and technology as social institutions, internally organized to produce knowledge and know-how, externally linked to and embedded in society at large. This whole dimension is missing from conventional science education, where the sciences are presented as if they had no social context, no social influence, and were of no social concern.

The sociological approach opens up into a wide area of study. It can show that individuals are trained in, and achieve scientific significance through, their institutions (universities, learned societies, invisible colleges of researchers, industrial laboratories, and so on). It shows that these institutions perform a variety of functions in society at large, as sources of knowledge and expertise, as educational media, as segments of governmental or corporate power, and so on. It brings politics and economics into the picture, showing how science is no longer a relatively self-contained activity but is being incorporated into the polity and the economy. All these are important matters for explanation and analysis in STS education.

From an ideological point of view—and STS education is not really without its ideology, even though we try to reduce it coolly to a rationale—the matters that arise in the course of a sociological approach provide valuable ammunition against the scientistic and technocratic attitudes that are taken for granted by most scientists and technologists. It puts social values and social consequences at the top of the criteria of relevance, and debunks the notion that there must be a technical fix for every social problem. This is a perception that most of us, as STS educators, would regard as essential nowadays for the thoughtful citizen, whether actively engaged in science or merely nourished by its fruits.

But the sociological approach also has quite serious disadvantages. Strangely enough, although quite young children are accomplished social actors, and adults discuss endlessly the policies and practices of particular institutions, people seem to find a more general analysis of social structures very difficult and abstract. The sociological mode of thought seems to be just as rare as the philosophical mode, and no more attractive to the average student at school or university. Indeed, it seems easier to grasp the philosophical notion of a theory than the sociological notion of a paradigm. Most people find the discovery of disconfirming evidence a more convincing explanation for the closure of a scientific controversy than the defeat of a coalition of interested parties.

It is also unhealthy for STS education to replace naive scientific realism with the naive relativism currently advocated by some schools of the sociology of science. Total skepticism is an irrefutable but completely barren epistemological standpoint, which is used much too freely as a refuge by lazy polemicists. Sociological analysis is the key to a proper understanding of belief in science, but it does not make obsolete some very convincing philosophical arguments that go back to the ancient Greeks.

The "Problematic" Approach

The principal goal of many exponents of STS education is to offer enlightenment on the great problems of our times, problems such as the destruction of the natural environment, overpopulation, endemic disease, poverty, and war. These problems are so interrelated that they are often described as components of a single complex world *problematique*. It is widely argued that the grave situation in which humanity now finds itself is due to the misapplication of science and technology in the past, and that the only hope for the future is to use science wisely to put matters right.

It is not necessary to buy the whole of this argument to appreciate its force as a rationale for general STS awareness. It also suggests a major line of approach to STS education. All that is required is to teach straightforwardly about various aspects of the *problematique,* latching onto the concern that those various aspects arouse among thoughtful students. All the points that one might make about the social role of science will inevitably arise, and can be appreciated all the better for being presented in a realistic context.

The great virtue of this approach is that it attempts to deal directly with the world as it is, representing science, technology, and society in contemporary concrete terms. Students and teachers can be enthused by a sense of personal commitment to a common cause, although it is obviously unwise to overemphasize purely subjective attitudes of sympathy and concern. Such an approach cannot avoid questioning the values that govern the pursuit of knowledge and of technical novelty, which are central to the whole STS theme.

It is not surprising that a great deal of STS education takes this general line. But this approach, too, has its defects. The most obvious is, of course, that the whole enterprise easily becomes a mere adjunct to a political or religious ideology. The evils of the *problematique* are attributed to capitalism, or socialism, or imperialism, or modernism, or reactionary fundamentalism, or what have you, which can only be combated by large doses of some other nostrum. Whatever the place of doctrinaire instruction in formal education, this is definitely not a suitable context for a dispassionate analysis of the social role of science and technology in the modern world.

Even when it does not get trapped in doctrine, the "problematic" approach often leads away from science and technology into a variety of other areas and aspects of society. The issues that cluster around the *problematique* are as much matters of law, economics, religion, and power politics, as they are of science.

Indeed, it is seldom that one finds within this whole complex of argument and opinion any serious analysis of the nature of science itself. All too often, it is taken as a "black box," either for solving problems or for generating irrelevant or inappropriate knowledge. Paradoxically, those who attribute the troubles of the world to science are as much bemused by its authority and power as the technocrats who believe that science-based technologies are the primary source of all that is good in life. It is the responsibility of STS education to explicate the false antinomy between science and antiscience by showing that science, too, is a fallible, fumbling human institution, which we use for good or ill, as best we can.

THE MANY FACETS OF *STS* EDUCATION

STS education has a variety of approaches. The approach through *relevance* leads into technology, but not deeply into social questions. The *vocational* approach raises these questions, but often answers them too technocratically. The *transdisciplinary* approach emphasizes the unity of the sciences and their associated technologies, but may exaggerate the power of science and technology in dealing with social problems. STS education is shallow without a *historical* dimension, but this easily becomes excessively academic. A *philosophical* approach can give some idea of the nature of science, but only at the most elementary level. The *sociology* of science can explain the nature and role of science as a social institution, but is often unhelpfully skeptical in its conclusions. Finally, direct consideration of the world *problematique* raises many of the most important STS issues, but says little about how science and technology actually work, for good or for ill.

The message of this brief survey of the many different approaches to STS education must now be clear: there is no single royal road. Each approach has its particular rationale, its thematic virtues, and its pedagogic advantages. Each also has quite severe limitations, in practice and in principle.

A more detailed analysis shows, however, that the different approaches have complementary strengths and weaknesses. The model of science that is missing from the approach that uses the world *problematique* can be derived from philosophy and sociology. All the other approaches can be enriched by historical examples. The approach through relevance can be given a strong emphasis on the *problematique,* and so on.

I am not arguing here for a shallow eclecticism—a smorgasbord of trendy topics served out to the students without any information about how they might be related. STS education is too complex and diverse to be grasped or presented as a completely consistent and thoroughly integrated academic discipline. But we can give our students some idea of the connections, including contradictions, between its various aspects. One might even think of it as a gemstone, whose facets are less interesting than the edges and corners where they meet.

I am sure that what I have said here is now widely accepted in the STS movement. Every STS teacher has to choose the messages and methods that are appropriate to the educational context in which he or she is working. Sometimes these choices seem a little narrow, a little vague, a little unimaginative, or a little doctrinaire. What I am trying to do is to remind STS teachers of the variety and richness of the fare from which their choices might be made.

Chapter 4

STS in the School Curriculum: A Movement Overtaken by History?

David Layton

A few years back I had the opportunity to work for a spell in Australia, at Monash University, where there is a strong science education group. In the course of that visit I had some discussions on the nature of curriculum change with Dick Selleck, who occupies a chair in the history of education there.

It was Selleck's view, based on historical studies of curriculum change and offered provocatively to deflate overconfident advocates of rational planning, that if you wanted a curriculum innovation to take off, you should never define it too precisely. By allowing a penumbra of ambiguity around the boundary of the innovation, interest groups of many kinds would see within it ways in which it might serve their different ends. As a result, they would lend it their support. Once rolling, it was difficult for them to withdraw from the bandwagon, even though they might not like the direction of travel. Having invested effort, there was always the hope that if they stayed on board, they might still influence the direction of change toward their preferred goal. Of course, as we now know, the differences between the intended, the enacted, and the achieved curriculum can be considerable; what students carry away from attempts at curriculum change can be very different from the outcomes that the original proponents had envisaged.

I think Selleck would find some support for his thesis in the STS movement. It has never been a precisely defined innovation and it has been impelled by an intriguing mixture of motives, yielding different curriculum manifestations. Some of its roots can be traced back a long way; the quests for a better understanding of the nature of science as an intellectual activity (the social studies of science) and for a deeper knowledge of the interactions of science and society (science policy studies) have been pursued by academics for much of this century, if not longer (Spiegel-Rösing, 1977). But in broad terms, STS education in schools is very much a product of the late

1970s and 1980s. Although we have no right to expect history to fall neatly into the Procrustean beds of centuries and decades, on this occasion there is at least an element of truth in the popular view that the 1960s was the decade of single-subject sciences, the 1970s was the decade of integrated science, and the 1980s was the decade of STS.

This chapter explores the development of STS education, first by examining six specific features that have characterized STS education, and then by showing why recent forces within society now challenge the STS education movement.

SOME FEATURES OF SCHOOL *STS* EDUCATION

Time is too short, and this is not the occasion, for a well-worked and detailed history of the school STS movement. Instead, I would like to take a brief global view of school STS education to identify some features that any history would need to address and attempt to explain. There are at least six characteristics of the movement that strike me as interesting and thought-provoking.

Origins

First, the school movement has had separate but roughly simultaneous origins in secondary education in a number of countries, mostly in Europe and North America.

In the United Kingdom the first major curriculum in the school STS field dates from 1976, when the Association for Science Education's *Science in Society* project materials were being published. Others were to follow, notably the SISCON (Science in a Social Context)-in-Schools project, directed by Joan Solomon (1983). Independently in the United States, within Project Synthesis (the major endeavor funded by the National Science Foundation in 1978 to derive recommendations for future action in science education from the voluminous products of evaluatory reviews conducted in the mid–1970s), an STS focus group was identifying areas of concern and urging the inclusion of STS materials in science textbooks. Interestingly, in the United States, a separate push for STS education came from social studies teachers concerned with the issue of good citizenship. In 1977 the National Council for the Social Studies gave autonomous status to its Science and Society Advisory Committee and its journal *Social Education* devoted an issue in 1979 to science and technology for a global society. Somewhat earlier in Canada, Glen Aikenhead and Reg Fleming had produced the first draft of *Science: A Way of Knowing* in 1973. Based on classroom research, the book was published in 1975.

Simultaneous developments in the Netherlands took a different turn, being focused on one subject, physics. Starting in 1972, the PLON project began to modernize and update the school physics curriculum by broadening its goals and later incorporating material on the interactions among physics, technology, and society.

The isolation of these new projects came to an end when international networks of various kinds were formed during the early 1980s. In the spring of 1980, the Committee on the Teaching of Science of the International Council of Scientific Unions held a seminar at Malvern, England, attended by representatives from 20 countries. Of the 46 participants, 33 came from Europe and North America. It was at this seminar that the seeds were sown for the Bangalore Conference 5 years later on Science and Technology Education and Future Human Needs.

A different network arose from the International Organization for Science and Technology Education (IOSTE) second symposium on World Trends in Science and Technology Education at Nottingham in 1982. This has a most valuable resource in an *STS Research Network Missive,* produced regularly by Glen Aikenhead from the University of Saskatchewan.

There have been other developments for breaching the insularity of STS developments, such as UNESCO's International Network for Information in Science and Technology Education (INISTE), inaugurated in 1984. Despite these efforts to build a community of STS teachers and researchers, a conceptual framework for school STS studies has proved an elusive goal. Susan Cozzens (1990) has gone so far as to assert that the STS movement has no theoretical core and its ideal, an integrated and accessible body of knowledge to inform the STS movement, is still awaited.

Location

A second striking feature of school STS education, viewed historically, is that it has been predominantly located in secondary education in industrialized, "developed" countries. With rare exceptions, STS did not feature as a slogan in science education activities in developing countries during the 1970s and earlier. (I use the term "developing countries" here to include those with a relatively low income per capita and with economies that are largely nonindustrial, for example, rural agricultural.) Some leading industrialized nations, such as Japan with its reformed science curricula, have paid little attention to STS issues in formal education. The same is true of some of the newly industrialized countries. Interestingly, one of the few Japanese STS projects is based on the family as the educational unit and is strongly oriented to moral and values education within a Confucian frame (Ogawa, 1989).

There is no need to carry out a careful quantitative analysis of the membership list of IOSTE's STS Research Network to appreciate that the balance is tipped heavily on the side of the industrialized nations, although interpretation of this bias needs to be cautious. It could be, at least in part, no more than a reflection of organization and communication mechanisms for science education in these countries. What is at least as striking as the country imbalance is the different interpretation given to "science, technology, and society" by those science educators from developing countries who had initiated work in this area.

At the Malvern Seminar in 1980, 7 of the 46 participants came from developing countries. It would be true to say that they were not impelled by the need to incorporate in their science teaching a more authentic view of science in its complex social context and to dispel mythical textbook images of the activity. Of course, they acknowledged the importance of these needs and they were unyielding opponents of scientism and technocracy. In many of their contexts, however, the prime function of science education was to assist in the battle to survive, to achieve self-sufficiency, to reduce dependency, and to raise rapidly the quality of daily life. The examples of school STS developments they described were ones such as the San Salvador project (a community science education project in an impoverished small fishing village) undertaken by the Institute for Science and Mathematics Education Development (ISMED) of the University of the Philippines, and designed to identify and satisfy community needs, for example, safe drinking water, adequate sanitation, and improved productivity in methods of fishing (Jesuitas, 1988). Another was a Mexican project, again aimed at community development, with a focus on the health care of children under five, involving collaboration between healthcare workers and teachers. School students (and dropouts) were recruited to take a census of the children, and subsequently to register vaccinations, prepare and distribute certain dietary essentials, and monitor and record progress.

It would seem that school STS education has been markedly influenced by the cultural context in which it has been located and that the functions of STS education in developing countries are not identical to those in many highly industrialized nations. A crude distinction might be between "quality control" on standards of living that are already high, in global terms, for the developed world, and "practical action for survival and self-sufficiency" elsewhere.

Target Audience

A third feature of school STS developments in the industrialized countries is that they originated very much at the higher end of secondary education. In the United Kingdom, both the *Science in Society* project and the *SISCON-in-Schools* project were aimed at students aged 16 to 18 years. The same was true of early United States developments such as the Biological Sciences Curriculum Study program, inaugurated in 1980, called *Innovations: The Social Consequences of Science and Technology* (McConnell, 1982). Canadian, Australian, and Dutch projects also were focused on those aged 15 and older. It is only comparatively recently that attempts have been made to initiate STS teaching in the lower secondary school and beyond. The first Science and Technology in Society (SATIS) units for the age range 8 to 14 years were published by the Association for Science Education in the United Kingdom in 1992, with more to follow in 1993.

Primary School Science

This relates directly to the fourth feature, which is that STS has not been part of the familiar rhetoric of primary school science education to any extent in industrialized countries. The emphasis instead has been on firsthand exploration of natural phenomena as a source of conceptual understanding, and science as a stimulus to language development, a motivator for reading skills acquisition, and a source of creative responses expressed in both the visual and nonvisual arts and through inventiveness in technological design and construction (Association for Science Education, 1991). The potential for STS education at the primary level has clearly existed because much good primary school science teaching has been intimately related to technological activities and to the world outside school. However, there have been few attempts until recently to explore what might constitute progression in STS learning as a planned and continuous process throughout the years of compulsory schooling. The exploration of what STS education might look like at the primary level is also a matter of significance for developing countries where, for the majority of school-aged children, their primary education is all they will ever experience in the way of formal education.

School and Higher Education STS

The fifth feature is that the school STS movement has had its own internal dynamics. To a large extent it developed separately from and, in its origins, derived little guidance from the Social Studies of Science movement in higher education. It is true that the latter took off before the school movement. In the United States, the STS programs of universities such as Cornell and Penn State were in place by the end of the 1960s. In the United Kingdom, newly created university departments (for example, Liberal Studies in Science at Manchester) and units (the Science Policy Research Unit at Sussex and the Science Studies Unit at Edinburgh) were products of the mid–1960s. It is also true that there were occasional higher education innovators who went into the schools in an attempt to colonize them with their particular brand of STS. Likewise, there were school-based entrepreneurs who carried out raids into Social Studies of Science territory in higher education to capture the bits that supported their own STS prejudices. In the early years, however, the relations between the school and the higher education STS movements were attenuated. The SISCON-in-Schools project marked the first major United Kingdom attempt to relate the developments. Another landmark that inaugurated a more fruitful relationship was John Ziman's (1980) book, *Teaching and Learning about Science and Society*.

Proportion of STS Content

The last feature to which I wish to draw attention is the disagreement that still exists about how much STS to have in the school science curriculum. For some, such as ICSU's Committee on the Teaching of Science, a 10% to 20% infusion, a carefully controlled leavening of academic science, is what is needed. For others, STS *is* science education. The National Science Teachers Association in the United States views STS as "the teaching and learning of science in the context of human experience" and asserts that it "represents an appropriate science education context for all learners" (National Science Teachers Association [NSTA], 1990). However, Project 2061 of the American Association for the Advancement of Science, in its report *Science for All Americans* (American Association for the Advancement of Science [AAAS], 1989), makes little direct reference to STS. Its project director, F. James Rutherford, in asking, "Will STS be just another passing fancy?" has argued that STS needs to be seen as a designation for a shared interest in changing schooling in ways that reflect the centrality of science and technology in today's world (Rutherford, 1988). This is somewhat less than wholehearted support for STS as the leading edge of reform in science education today. The debate continues and some of the tensions afflicting STS at the point of implementation and incorporation in the school curriculum have been well described by Peter Fensham (1988, 1992) and Joan Solomon (1988b).

INTERPRETATIONS AND EXPLANATIONS

What I have been outlining so far are some, though by no means all, of the matters that any history of school STS education would need to address and account for. To interpret and attempt to explain them, both microlevel (classroom) and macrolevel (societal) considerations need to be invoked.

At the microlevel, unquestionably some STS developments in secondary education were driven by student dissatisfaction with the academic science curriculum they were being offered. There was also evidence, not always well-founded, of poor and declining performances by students in learning science. It was estimated that in the 15 years following Sputnik 1 in 1957, the United States had invested 5 billion dollars in programs to improve school science education and accountability concerns were running strongly. In the United Kingdom, Nuffield science courses were being criticized for their internalist orientation: they appeared to ignore the application of scientific principles to industrial and everyday situations. There was much talk of science being not just for inquiring minds, but for action and for citizens as well (for example, Lewis, 1985).

Many early STS developments were a response to this vein of criticism. Their significance was reinforced by macrolevel influences. The post–World War II population bulge was entering school systems and also, in ex-colonial and now independent countries, educational services were expanding rapidly. Not only more but more diverse students were being taught science as previously elitist educational provisions were adapted to become mass systems. The need was for science teaching materials that had greater motivational value and more immediate relevance to students.

Overlaying all this were influences from the political and economic context. The 1970s was the period of the Vietnam War, the rise of green politics, greater environmental awareness, and the oil crisis. There came into focus a long catalogue of issues relating to the use and misuse of science and technology and to their complex relationships in the research and development process within particular cultural contexts. A movement for Social Responsibility in Science became active. All of this helped to bring STS into prominence on educational agendas. Indeed, in the United Kingdom we had it on the very highest authority, as early as 1971, that we should be moving our science education in the direction of STS. At that time, Mrs. Margaret Thatcher was secretary of state for education and in her Ford Lecture to Youth she asserted that "school science ought to go beyond the Bunsen burners, the test tubes and the retorts, important as those are, and examine also the relationship between science and society" (Thatcher,

1971, p. 6). With uncharacteristic and unrepeated generosity to universities, she went on to say:

> I heard the other day about a development in one of our university departments of education. Science as an inward-looking subject with a formal structure is very necessary for future scientists, but what the rest of us need to know is how science is changing the economic, political and ethical problems of our time—what might be called the external relations of science. (Thatcher, 1971, p. 7)

I like to think this derived from a visit one of her speech writers had made to my own department at Leeds where we had already begun to prepare science graduates for teaching about the relationships among science, technology, and society.

Whereas the emergence of school STS is understandable in these terms, its absence from the lower levels of secondary education and from primary education is more difficult to explain. Pragmatic considerations probably played some part. Frequently the opportunity to introduce a new course was more readily available in the higher level secondary education curriculum. In the earlier years of secondary education, the constraints of external examinations could be severe. In England and Wales the curriculum space allocated to General Studies in the sixth form was an obvious location for STS courses and much of the earlier material was developed and trialed there.

There was also an assumption, rarely made explicit, that the kinds of issues on which the work was based required a social and intellectual maturity, and a knowledge of the world, that younger children could not display. To the extent that STS was preparing future citizens for decision making in a democracy, STS seemed to entail "an attitude of attentiveness; skills in gaining and using relevant knowledge; values of which one is aware and to which one is committed; and the ability to turn attitudes, skills and values into actions" (McConnell, 1982, p. 13). The cognitive complexity of STS issues was seen as formidable and the upper reaches of secondary education seemed the appropriate place to start. The belief that younger children were unable to engage profitably with these issues (and that therefore teaching approaches and materials could not be designed for them) was invalidated by empirical research. Subsequent professional experience of integrating STS issues into the science curriculum for junior secondary school students and primary school students (for example, Nganunu, 1988; Yager, Blunck, Binadja, McComas, & Penick, 1988) suggests that progress is possible on this front. However, formal evaluations of outcomes are scarce.

CHANGING CONTEXTS YET AGAIN

The history of school science education provides many reminders that prolonged contextual stability is the exception rather than the rule. With this in mind, I want now to look forward, rather than backward, in order to speculate on a possible future for STS while keeping the historical origins and development of STS education in mind.

What then are the broad changes in the context for school science to which we need to be sensitive in making judgments about the way forward for STS? There is, first, a considerable disenchantment among employers who have received the products of secondary schools with what they discern as a prevailing academicism. This manifests itself in various ways.

In Western industrialized nations, disquiet about economic competition from Pacific Rim countries is often related to a failure to educate students adequately in science. In 1986, the director of the U.S. National Science Foundation, when addressing the first National Conference on Technological Literacy, made his case for improving science education programs almost entirely in terms of their contribution to meeting the economic challenge from abroad (Bloch, 1986). In less industrialized countries, the criticism has been that school science does not lead to personal and collective empowerment in relation to urgent practical problems. Too often school science remains an arid and compartmentalized academic activity.

At the same time, one alternative to the traditional academic curriculum (diversification to include more practical and occupationally related subjects at the secondary level) has yielded few of the advantages claimed for it. There is little evidence of greater equity, in the sense of less-privileged socioeconomic groups achieving more, or of improved labor market outcomes such as reduced youth unemployment (Middleton, Ziderman, & Van Adams, 1991).

Simultaneously, the very terminology being used to describe the problem of unemployment is itself being reinterpreted. "Vocationalism" can no longer be understood as "narrow occupationalism." Profound changes have been associated with the globalization of commerce and industry and with the impact of modern technologies on the skill requirements of manufacturing processes and of commercial services. These have led to the identification of general competencies and foundation skills that are now essential prerequisites for employment. According to a report of the Secretary's Commission on Achieving Necessary Skills (SCANS) established by the U.S. secretary of labor, these need to be injected "into every nook and cranny of the school curriculum" (Secretary's Commission on Achieving Necessary Skills, [SCANS] 1991, p. ix).

Expressed differently, both the academic and the vocational are being required to change. The former is becoming more related to applications in real-life situations and the latter is moving from specificity to generality. In so doing they converge on a new curriculum model that has practical capability at its heart.

From the standpoint of STS, the most significant aspect of this change is the emergence of technology as a component of the curriculum of general education and an autonomous coequal to, and not a mere adjunct of, science. Though recent, this is now a worldwide phenomenon with powerful support from governments and industry. Technological literacy has become a slogan of the 1990s, with international conferences and a burgeoning literature devoted to it (for example, Blandow & Dyrenfurth, 1992; Dyrenfurth & Kozak, 1991; Layton 1993a, 1993b).

It is true that much of the drive for technology education has come from those who see education as instrumental to economic competitiveness and wealth creation. However, there has been significant support from other quarters, not least from those concerned about the control of technology and the compatibility of technological innovations with sustainable development and environmental quality. Of great importance also has been our increased understanding of the nature of technological practice and of technological knowledge. History, philosophy, and sociology of technology, as academic fields of inquiry, have developed rapidly in the second half of the 20th century, much as the history, philosophy, and sociology of science did in the first half. As a result, it is no longer tenable to view technology as "merely applied science." The hierarchical dependency model of the relationship between science and technology has been replaced by one that reflects more accurately the symbiotic and synergistic interplay between equal partners.

In passing, it may be worth noting a contemporary paradox. As their symbiotic relationship draws science and technology closer together in research and development activities, their institutional manifestations are moving towards separatism in many contexts. To give just one example, in the United Kingdom the quest by engineering institutions for higher status and for political and financial autonomy has led to the establishment of a Royal Academy of Engineering, intended to be comparable to the long established Royal Society (for scientists). At the microlevel of the classroom, the emergence of technology education as an autonomous component of general education might be seen as a parallel development influenced by the recognition that technological activity is distinctive and different from science.

IMPLICATIONS FOR SCHOOL *STS* EDUCATION: SOME QUESTIONS

It is possible that the school STS movement has already been overtaken by history, or at least challenged by it, in at least three different ways.

First, much of what has taken place under the banner of STS has been a product of developments in the late 1960s and the 1970s in Western Europe and North America by a science education community whose understanding of "technology" did not seriously challenge its equation with "applied science." School STS was conceptualized before the academic study of technology (as opposed to "craft" or "industrial arts") had begun to be incorporated into the school curriculum. In Britain, and perhaps to a lesser extent North America, the situation is changing. Technology has now achieved a central position in the core of the new British National Curriculum and this status will certainly have an influence on the school curriculum. I have argued (Layton, 1988) that STS education must pay serious attention to how technology is involved in an STS curriculum. In the past, technology has frequently been subsumed under science. Any recognition of technology as an equal of science raises issues about the uniqueness of technological knowledge and how it is structurally oriented toward praxis (Layton, 1991). A comparison between science and technology establishes the value-laden nature of every phase of technological activity. It follows, therefore, that moral education is an inescapable consequence of the revaluation of the T in STS.

If we were to have a science curriculum in education with a focus on authentic science and an understanding of science in its social contexts, and if we were to add to that a separate technology curriculum that emphasized problem solving and inventions in the practical world, then what would be the role of STS? Insofar as the science and technology curricula need supplementation, it would probably come from the field of practical ethics and moral education. Was STS a coalition of its times and is it now in need of reformation?

Second, if indeed, as has been argued above, a change is taking place in the overall mission of schooling in the direction of "privileging the practical" (that is, relating students' experience more closely to the acquisition of practical capability in the world outside school) what does this mean for STS? Joan Solomon (1993) has identified four approaches to STS education that she designates as (1) *cultural* (a better understanding of the nature of science and technology and of their relations to society); (2) *political education for action* (educating future citizens for appropriate political action on controversial issues related to science or technology); (3) *interdisciplinary issue-based* (focusing on specific contemporary problems); and (4) *vocational or technocratic* (STS for technocrats and the economy). The debate is sometimes simplified by asking whether the end of STS is scholarship or activism. Clearly, if the educational context for school STS is changing in the direction suggested, the most appropriate versions of STS will be those that relate most directly to practical action, that is, (2) and (4).

There is, however, a note of reservation to sound in relation to (2), at least. The justification for STS in terms of improved citizen participation in decision making on scientific or technological issues confronting society is appealing, but in practice could well lead to frustration unless corresponding improvements in institutional mechanisms for participation become available. STS education as an antidote to citizen powerlessness in the face of technological change is a concept overloaded with responsibilities that schools alone are not equipped to discharge. And participation is itself an ambiguous term with meanings that range from a high degree of control through to placation and mere consultation (Arnstein, 1977). A move from "cultural STS" to "practical action STS," while holding out many attractions and, on the face of it, congruent with changes in schooling already underway, would unquestionably bring problems in its wake and would test the limits to which the development of "practical capability" could be pushed.

Third, as the needs of the developing world become more prominent on educational agendas, following the World Conference on Education for All at Jomtien in 1990 and its commitment to universal access to primary education by 2000, do we need to look again at school STS and ask to what extent it is premised on a model of participatory democracy and on assumptions about student-teacher relationships that do not transfer readily to all other contexts?

The problem is one of cultural diversity, specifically, how STS education affects cultural diversity. One role, among others, of science education in developing countries is the formation and maintenance of national cultures. The insinuation of values and practices foreign to, or destructive of, that culture should not be a consequence of adopting STS as a model for the school science curriculum. Yet, as with technologies, curricula bear within them the imprints of their authors and can act as social genes carrying social relations from one society to another. Conceived and honed in the Western world, STS may need to take a very different form in systems elsewhere. There are, for example, marked national differences in the ways in which values education and moral education are approached, and these are inescapably core components of any study of the interactions of science, technology, and society (Cummings, Gopinathan, & Tomoda, 1988). The African and Indian perspectives on STS education described by Olugbemira Jegede (Chapter 12) and Anita Rampal (Chapter 13) illustrate such national differences and underscore the critical importance of sensitivity toward cultural diversity.

It may be that school STS education has within it the resources and flexibility to survive these and other challenges and adapt successfully to the various contextual changes that confront it. Alternatively, STS could turn out to be like geography; critics argue that there is no such subject because, on close inspection, all geographers are found to be economists, historians, geologists, or mathematicians. The difference is that no one discovered this fact before geography became institutionalized. Though influential, STS as yet can hardly claim that status. However, as Paul Durbin (1991) has argued, it does provide a forum for cross-disciplinary (hopefully, transdisciplinary) discourse that so far is unique. Until better structures or conceptual apparatus come along, STS remains in business.

Part II

STS in Science Classrooms

Part I described the general features of STS education for school science and posed challenges to educators. One challenging feature is the diversity of STS. Part II takes up this challenge by exploring how STS "flows in so many different channels," to use John Ziman's metaphor from Chapter 3. In North America, the channels of STS comprise science education, technology education, social studies, and general education (Gilliom, Helgeson, & Zuga, 1991, 1992). In Europe, and in this book, STS education primarily assumes a science orientation that itself abounds with diversity.

To help understand this diversity, Glen Aikenhead (Chapter 5) explores four key questions: What are the goals for teaching science through STS? What should be taught? How should the science and STS content be integrated? And how can we design STS instruction? He describes several conceptual frameworks that categorize different types of STS curricula and instruction.

Each of the ensuing chapters in Part II charts a different channel of STS teaching (to continue John Ziman's metaphor). In Chapter 6, Bill Williams, based on his wealth of experience, describes the grassroots experience of United States science teachers. He revisits some key projects and expresses personal concerns. His keen support of innovative teachers underscores the importance of their participation in developing the courses they will teach.

The British experience with STS science is captured by Andrew Hunt (Chapter 7) as he outlines the development of programs that have produced, and continue to produce, a wide variety of STS teaching materials. His personal familiarity with teaching all of the materials ensures a realistic account of those developments.

A different channel is explored in Chapter 8 by Herb Their and Barbara Nagle, where they describe a model for issues-oriented STS teaching, a response to David Layton's challenge concerning the empowerment of citizens. A unique feature of their model is the way content naturally emerges from industry and the everyday world.

William Carlsen and his colleagues analyze STS text materials that teachers might use (Chapter 9). Their chapter asks: Can we use STS materials in our classroom without being used by the ideological messages embedded in those materials? The authors draw on their research on an STS chemistry textbook and its instruction methods. The chapter demonstrates how teachers need to be vigilant to implicit ideological messages.

Although Part II focuses attention on STS approaches to science teaching, STS teaching is illustrated in the other parts of the book as well.

Chapter 5

What is STS Science Teaching?

Glen Aikenhead

David Layton (Chapter 4) described the STS movement as having "an intriguing mixture of motives." This mixture has produced an array of STS curricula (Cheek, 1992c) and a variety of approaches to STS outlined in Chapter 3 by John Ziman. It is interesting, and even reassuring, to read in David Layton's description the positive consequences to an *imprecisely* defined innovation. The present chapter clarifies this imprecision by providing several frameworks that help to make sense out of the "mixture of motives" in STS science teaching.

When we are actually teaching in a classroom, it is difficult to separate curriculum from instruction. But when we reflect on curriculum and instruction, it is advantageous to separate the two in our minds. The *instructional* features of STS science teaching (the methods and strategies for how to be an effective STS science teacher) have been thoroughly described by Joan Solomon (1993) and myself (Aikenhead, 1988b, 1991b). The *curricular* features of STS science teaching are the focus of this chapter.

Four aspects of a curriculum will be explored to clarify the imprecision in STS science teaching:

1. Function—what are the goals for teaching science through STS?
2. Content—what should be taught?
3. Structure—how should the science and STS content be integrated?
4. Sequence—how can we design STS instruction?

This chapter does not offer a definitive answer to these questions, but it does sketch the territory that must be explored when we try to answer the questions for ourselves.

FUNCTION

When compared to the function of the traditional science curriculum, STS science represents a type of Kuhnian paradigm shift in goals. STS educa-

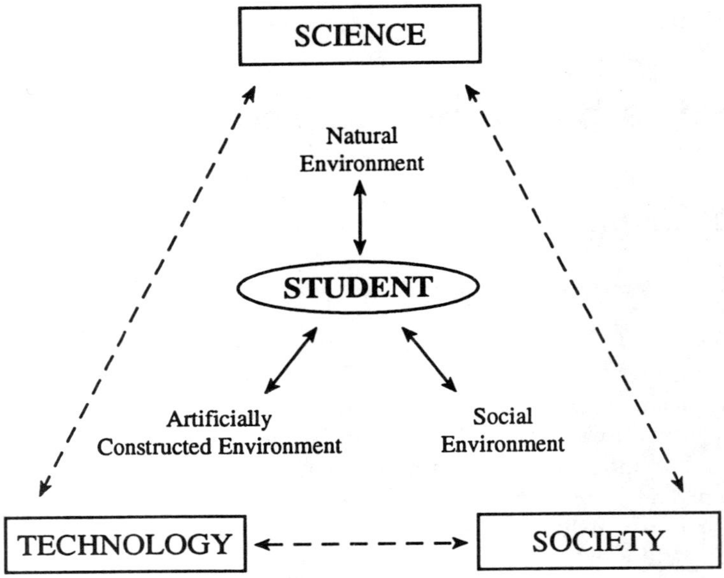

FIGURE 5.1 *The essence of STS education*

tion embraces the successes of the old paradigm but with a different worldview on science teaching. This worldview is described here.

Fundamentally, STS science teaching is student oriented, as contrasted with the scientist orientation of tradition science teaching. The student-oriented character of STS science is represented in Figure 5.1 by the central position given to the student.

Students strive to understand their everyday experiences. To do so, students make sense out of their social environment, their artificially constructed environment, and their natural environment. This sense-making is depicted in Figure 5.1 by the solid arrows between the student and the three different environments. Students *integrate* their personal understandings of their social, artificially constructed, and natural environments. This integration is represented in Figure 5.1 by the solid arrows *simultaneously* linking the student with each of these three environments. The study of the natural world we call science. The study of the artificially constructed world is technology. And society is the social milieu. Teaching science through science-technology-society (STS) refers to teaching about natural phenomena in a manner that embeds science in the technological and social environments of the student. This function is shown in Figure 5.1 by the broken arrows. The broken arrows attempt to superimpose a pedagogical structure that harmonizes with the solid arrows. In other words, STS instruction aims to help students make sense out of their everyday experiences, and does so in ways that support students' natural tendency to integrate their personal understandings of their social, technological, and natural environments.

The "SCIENCE" box in Figure 5.1 represents traditional science discipline content. In a traditional science curriculum, this science content is taught in isolation from "TECHNOLOGY" and "SOCIETY" (students' technological and social worlds). In an STS science curriculum, this science content is connected and integrated with the students' everyday worlds, and in a manner that mirrors students' natural efforts at making sense out of those worlds.

STS science is about making sense out of life today and for the future. But for what purpose? What are the goals of STS science education? Themes have emerged from Part I of this book. These themes are summarized here.

The inadequacies of traditional science education (the "saber-tooth" curriculum, Chapter 2) define one overarching theme. STS science is expected to reverse the existing negative trends in enrollments, achievement, and career choices. Specifically, STS science is expected to increase general interest in and public understanding of science, particularly for the bright creative students who are discouraged by a boring and irrelevant curriculum (Bondi, 1985; Eijkelhof, 1990; Kahle, 1988; Majumdar et al., 1991; Oxford University, 1989; Science Council of Canada, 1984; Sjøberg & Imsen, 1988; Solomon, 1988a, 1988b, 1993). The gender aspect of the problem is the focus of Part IV of this book.

STS science is also expected to fill a critical void in the traditional curriculum—the social responsibility in collective decision making on issues related to science and technology. Such issues require a harmonious mix of a scientific-technical elite with an informed attentive citizenry. Together both groups will need to make complex decisions that involve "the application of scientific knowledge, technological expertise, social understanding, and humane compassion" (Kranzberg, 1991, p. 238). The pervasive goal of social responsibility in collective decision making leads to numerous related goals: individual empowerment (Aikenhead, 1985a, 1985b; Fleming, 1989; Layton, 1986; Solomon, 1993); intellectual capabilities such as critical thinking, logical reasoning, creative problem solving, and decision making (Bybee, 1987); national and global citizenship, usually "democracy" or "stewardship" (Gilliom et al., 1991, 1992); socially responsible action by individuals (De Vore, 1992; Rubba, 1991); and an adroit work force for business and industry (Bondi, 1985; Hurd, 1989; Moore, 1991; Wirth, 1991). These goals emphasize an induction into a world increasingly shaped by science and technology, more than they support an induction into a scientific discipline.

STS science courses differ widely because of their different goals. On closer examination, however, this variation reflects differences in the *balance* among similar goals. In other words, most STS science courses harbor similar goals but give different priorities to different goals. The idea of balance is captured by the slogan "scientific literacy" (Hart, 1989; Roberts, 1983). Although "scientific literacy" provides an element of persuasion in rationalizing science programs (who can be against scientific literacy?), the term can be useful in defining a goal cluster.

For Bybee (1985b, p.85), the balance for STS science education is among three general goals:

1. "Acquisition of knowledge" (concepts within, and concepts about, science and technology) for personal matters, civic concerns, or cultural perspectives.
2. "Development of learning skills" (processes of scientific and technological inquiry) for information gathering, problem solving, and decision making.
3. "Development of values and ideas" (dealing with the interactions among science, technology, and society) for local issues, public policies, and global problems.

Another cluster of goals for STS science has been identified by Waks and Prakash (1985, pp. 108–114):

1. Cognitive competency—standardized knowledge and skills needed for reading and speaking accurately about STS issues; for example: conservation of energy in science means something different than it does in everyday use; controlling variables is essential to good experiments; research and development is a combination of science and technology.
2. Rational/academic—a grasp of the epistemology and sociology of science required for understanding the dynamics at play in STS issues; for example: scientific observations are theory-laden; scientific beliefs are reached by consensus making; epidemiology can have political dimensions.
3. Personal—students understand their everyday lives better; for example: money invested in insulation has benefits in cold climates; you are what you eat; giving careful attention to words on labels can save you problems.
4. Social action—students participate in responsible political action; for example: making consumer choices to affect global environments; writing letters to government or industries; participating in the resolution of a local issue.

All four goals may have a place within a single curriculum, but some goals will have higher priority than others. For example, the fourth goal—social action—is usually a high priority for environmental courses (Ruoba, 1991). This goal may have a low priority in some educational jurisdictions where communities discourage students from engaging in social action. An STS science course would likely embrace all four goals, though each goal with a different emphasis.

These and other goal clusters have guided curriculum development in STS science (Aikenhead, 1986; Cheek, 1992c; Fensham 1988; Hart, 1989; Northern Examination Board, 1988; Science Through STS Project, 1985; Solomon, 1993; Yager, 1992c).

The function of the old ("saber-tooth") science curriculum has been to prepare students for the next level of education and to teach correct answers (Roberts, 1988). These functions are not ignored in STS science, but they are not given as strong an emphasis. Therefore, an STS science curriculum addresses the needs of two groups: (1) future scientists and engineers (the elite), and (2) citizens who need intellectual empowerment to participate thoughtfully in their society (the attentive public; science for all).

A curriculum's goals are hollow without content. The next section considers the issue of content.

CONTENT

There is a marked difference between the content of university STS courses and the content of high school STS science courses. University courses invariably deal with science and technology policy, development, and/or discourse (Layton, Chapter 4; Lepkrowski, 1989; McGinn, 1991). The subject is abstract. On the other hand, high school STS courses position themselves among the concrete experiences of students. These courses provide high school students with a simplified, although intellectually honest, perspective on the human and social aspects of science.

The content of STS science will include *both* science content and STS content. Here I focus on STS content. In the section that follows, "Integrative Structure," I explore how this STS content can be integrated with science content.

Many educators, particularly in North America, conceive of STS science content as dealing mainly with social issues that connect science with a societal problem (Yager, 1992a). Rosenthal (1989) and Ziman (1984) remind us, however, that there are two types of social issues in STS science:

1. Social issues *external* to the scientific community ("science and society" topics; for example, energy conservation or pollution).
2. Social aspects of science—issues *internal* to the scientific community (the social epistemology of science; for example, the nature of scientific theories or the cold fusion controversy).

The relative importance of various social issues external to science was a topic researched by Piel (1981) in Project Synthesis, and by Bybee and Mau (1986) in their survey of international science educators.

STS science can also address social issues internal to science; for instance, its philosophy, its history, and its interpersonal dynamics. The social issues internal to science are delineated by Snow (1987). His "system of science" has three dimensions: the cognitive, the personal, and the sociological. The *cognitive* dimension includes experimental knowledge, hypotheses, theories, laws, and empirical observations, as well as the values that underlie them (for example, accuracy, coherence, fruitfulness, and parsimony). The *personal* dimension encompasses a scientist's social values that influence his or her research programs and nonempirical arguments. The *sociological* dimension incorporates community values, invisible colleges, credibility, journal publications, and other aspects of importance to the scientific community. In his book, *An Introduction to Science Studies,* Ziman (1984) provides a thorough and systematic treatment of STS content, both external and internal to the scientific community.

The full range of STS content in science education is made evident in Cheek's (1992c) thorough review of 15 STS projects, syllabi, guidelines, and major policy statements. He analyzed projects from all over the world, but gave special attention to the STS content proposed by Rubba and Wiesenmayer (1985a), Waks (1987a), Aikenhead (1986), and the Science Through STS Project (1985).

The science education community holds a variety of views concerning STS content. Nevertheless, a succinct definition of STS content is offered here. The definition attempts to encompass the full range of views held by science educators. STS content in a science education curriculum is comprised of an interaction between science and technology, or between science and society, and any one or combination of the following:

- A technological artifact, process, or expertise
- The interactions between technology and society
- A societal issue related to science or technology
- Social science content that sheds light on a societal issue related to science and technology

- A philosophical, historical, or social issue within the scientific or technological community.

This broad definition is used in the next section to describe the structure of STS curricula.

INTEGRATIVE STRUCTURE

Various STS science curricula embrace different goals and content due to different views about the nature of STS. To clarify STS education further, a descriptive scheme is proposed, "Categories of STS Science" (Table 5.1). It delineates the diversity in STS science in terms of the degree and manner in which STS content is integrated with traditional science content. In other words, the scheme deals with the integrative structure of STS science education.

Before describing the scheme, let me comment on its limitations. The scheme does not attempt to evaluate different approaches to STS science. Nor does it attempt to prescribe any particular set of goals, or goal priorities, as mentioned above. Moreover, the scheme does not address: teaching methods (for example, inquiry, problem solving, decision making), contexts for instruction (for example, local issues, public policies, global problems), and assumptions about how students learn (though constructivism predominates in STS science; Cheek, 1992c).

The proposed scheme, "Categories of STS Science" (Table 5.1), characterizes STS science in terms of:

- Content structure—the proportion of STS content compared with traditional science content, and the way the two are combined;
- Student evaluation—the relative emphasis given to STS versus traditional content. The description is an approximate indicator of relative emphasis, rather than a prescription for classroom practice.
- Concrete examples of STS science. Each category in Table 5.1 is illustrated by titles of published teaching materials for schools, along with the names of their publishers. Due to space limitations, the list of titles is incomplete.

A spectrum underlies the proposed scheme. The spectrum expresses the relative importance afforded STS content in a science course. At one end of the spectrum (category one), STS content is given lowest priority compared with traditional science content, while at the other end (category eight), it is given highest priority. The eight categories of the spectrum are: (1) motivation by STS content, (2) casual infusion of STS content, (3) purposeful infusion of STS content, (4) singular discipline through STS content, (5) science through STS content, (6) science along with STS content, (7) infusion of science into STS content, and (8) STS content. One can think of each category as a conveniently identified point along the spectrum. Although no particular category can be said to represent "true" STS science instruction, categories three to six represent views most often cited by STS science leaders. Table 5.1 was inspired by a table about technology education in an article by Fensham (1988).

In categories one to three of the proposed scheme (Table 5.1), the selection and sequence of science content follow traditional lines. Students are expected to view their natural environment with a theoretical scientific mind. However, in categories four to eight, the structure changes dramatically. The organization of science content follows a sequence dictated by the STS content itself. Students view the everyday world from their own commonsense point of view, and draw upon science content on a need-to-know basis (Aikenhead, 1992c). This type of organization is discussed in the following section, "Sequence."

Table 5.1 contains numerous details on each of the "Categories of STS Science" and provides a language for talking about STS curricula, classroom materials, and teaching. For instance, some science teachers may be described as showing resistance to moving further than category one, whereas others may be described as feeling comfortable with teaching category three. On the other hand, certain STS educators might be classified, for instance, as advocating category four STS science.

SEQUENCE

Teaching science through STS may be organized according to a traditional science syllabus (categories one to three in Table 5.1), or according to a natural sequence suggested by the STS content itself (categories four to eight in Table 5.1). Which approach is more effective?

Chapter 16 reviews the research into this question. One conclusion is particularly germane to the present discussion. The research generally indicates that STS materials are best organized in a sequence indicated by the arrow in Figure 5.2 (see page 57). The arrow maps out a sequence of teaching events described here.

STS instruction begins in the realm of society, represented by the box in Figure 5.2. A key question or problem is posed; for instance, "Should we be concerned about high voltage power lines in our community?" "How can we explain the conflicting scientific testimony in a newspaper

TABLE 5.1 *Categories of STS science*

1. MOTIVATION BY STS CONTENT

Traditional school science, plus a mention of STS content in order to make a lesson more interesting. (The low status given to sts content explains why this category is not normally taken seriously as STS instruction).

Students are not assessed on the STS content.

Examples: What many teachers do now to spice up the pure science content.

2. CASUAL INFUSION OF STS CONTENT

Traditional school science, plus a short study (about ½ to 2 hours in length) of STS content attached onto the science topic. The STS content does not follow cohesive themes.

Students are assessed mostly on pure science content and usually only superficially (such as memory work) on the STS content (for instance, 5% STS, 95% science).

Examples: Science and Technology in Society (SATIS, U.K.: Association for Science Education), *Consumer Science* (U.S.: Burgess), *Values in School Science* (U.S.: R. Brinckerhoff, Phillips Exeter Academy, Exeter, New Hampshire).

3. PURPOSEFUL INFUSION OF STS CONTENT

Traditional school science, plus a series of short studies (about ½ to 2 hours in length) of STS content integrated into science topics, in order to systematically explore the STS content. This content forms cohesive themes.

Students are assessed to some degree on their understanding of the STS content (for instance, 10% STS, 90% science).

Examples: Harvard Project Physics (U.S.: Holt, Rhinehart and Winston), *Science and Social Issues* (U.S.: Walch), *Science and Societal Issues* (U.S.: Iowa State University), *Nelson Chemistry* (Canada: Nelson), *Interactive Teaching Units for Chemsitry* (U.K.: Newcastle Polytechnic), *Science, Technology and Society, Block J* (U.S.: New York State Education). Three SATIS 16-19 modules (*What is Science? What is Technology? How Does Society Decide?* U.K.: Association for Science Education.)

4. SINGULAR DISCIPLINE THROUGH STS CONTENT

STS content serves as an organizer for the science content and its sequence. The science content is selected from one science discipline. A listing of pure science topics looks quite similar to a category 3 science course, though the sequence would be quite different.

Students are assessed on their understanding of the STS content (for instance, 20% STS, 80% science).

Examples: ChemCon (U.S.: American Chemical Society), the Dutch physics modules such as *Light Sources* and *Ionizing Radiation* (Netherlands: PLON, University of Utrecht, Physics Department), *Science and Society Teaching Units* (Canada, Toronto: Ontario Institute for Studies in Education), *Chemical Education for Public Understanding* (U.S.: Addison-Wesley), Science Teachers' Association of Victoria Physics Series (Parkville, Australia: STAV Publishing).

TABLE 5.1 *Continued*

5. SCIENCE THROUGH STS CONTENT

STS content serves as an organizer for the science content and its sequence. The science content is multidisciplinary, as dictated by the STS content. A listing of pure science topics looks like a selection of important science topics from a variety of traditional school science courses.

Students are assessed on their understanding of the STS content, but not as extensively as they are on the pure science content (for example, 30% STS, 70% science).

Examples: Logical Reasoning in Science and Technology (Toronto: Wiley of Canada), *Modular STS* (U.S.: Wausau, Wisconsin), *Global Science* (U.S.: Kendall/Hunt), the Dutch Environmental Project (Netherlands: NME-VO, University of Utrecht, Physics Department), *Salters' Science* Project (U.K., Helsington: University of York, Department of Chemsitry.)

6. SCIENCE ALONG WITH STS CONTENT

STS content is the focus of instruction. Relevant science content enriches this learning.

Students are assessed about equally on the STS and pure science content.

Examples: Exploring the Nature of Science (London: Blackie & Son), Society Environment and Energy Development Studies (SEEDS) modules (U.S.: Science Research Associates), *Science and Technology 11* (Canada, Victoria: BC Ministry of Education).

7. INFUSION OF SCIENCE INTO STS CONTENT

STS content is the focus of instruction. Relevant science content is mentioned, but not systematically taught. Emphasis may be given to broad scientific principles. (The materials classified as category 7 could be infused into a standard school science course, yielding a category 3 STS science course.)

Students are primarily assessed on the sts content, and only partially on pure science content (for example, 80% STS, 20% science).

Examples: Studies in a Social Context (SISCON) *in Schools* (U.K.: Association for Science Education), *Modular Courses in Technology* (U.K.: Schools Council), *Science: A Way of Knowing* (Canada: University of Saskatchewan, Department of Curriculum Studies), *Science, Technology and Society* (Australia: Jacaranda Press), *Creative Role Playing Exercises in Science and Technology* (U.S.: Social Science Education Consortium), *Issues for Today* (Canada: GLC Silver Burdett), *Interactions in Science and Society*—a video series (U.S.: Agency for Instructional Technology), *Perspectives in Science*—a video series (Canada: National Film Board of Canada).

8. STS CONTENTS

A major technology or social issue is studied. Science content is mentioned but only to indicate an existing link to science. (The materials classified as category 8 could be infused into a standard school science course, yielding a category 3 STS science course.)

Students are not assessed on pure science content to any appreciable degree.

Examples: Science and Society (U.K.: Association for Science Education), *Innovations: The Social Consequences of Science and Technology* program (U.S.: BSCS), *Preparing for Tomorrow's World* (U.S.: Sopris West, Inc.), *Values and Biology* (U.S.: Walch), and in general, technology design courses and social studies modules

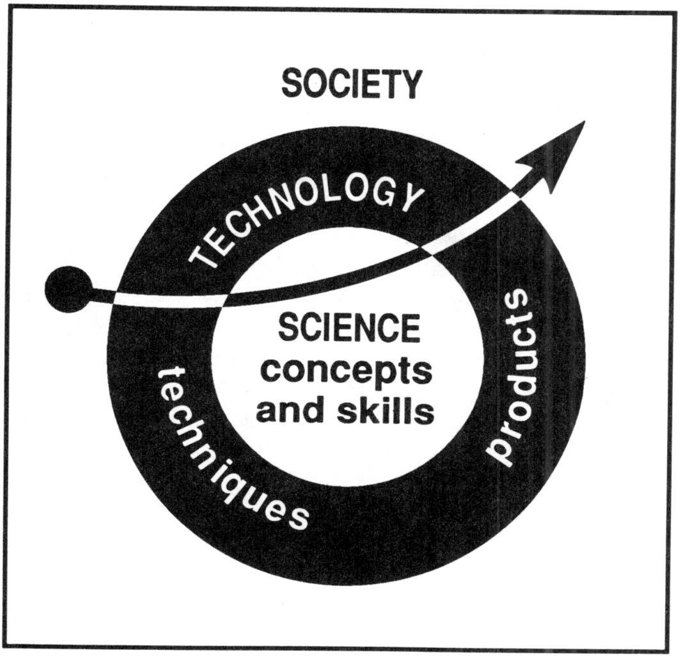

FIGURE 5.2 *A sequence for STS science teaching*

article?" "How is evidence of drunk driving challenged in a court of law?" or "For what purposes are we using different kinds of light sources?"

In order to understand a societal question or problem, there is usually some technology to examine, even at a superficial level. The domain of technology is represented by the black donut in Figure 5.2. Technology is primarily concerned with developing knowledge and designing processes, in response to human needs and societal problems (Collingridge, 1989; Harrison, 1980; Snow, 1987). Hence, social issues are almost always related to technology. Students are affected far more by their technological world than by their scientific world.

As mentioned above, a societal question or problem (the beginning of the arrow in Figure 5.2) creates the need to know certain technological knowledge (the donut area in Figure 5.2). But both create the need to know some science content (the central circle in Figure 5.2); for example, "What is the difference between epidemiological and etiological investigations in science reported in the newspaper?" "How does ethanol diffuse from the circulatory system into the respiratory system?" or "What is light?" This science content will help students understand the technology and the societal issue.

The sequence of instruction suggested by the arrow in Figure 5.2 begins in the domain of society, moves through the domains of technology and traditional science, and then out again to technology. There is an advantage to revisiting the technology that students had earlier studied (for example, high voltage power lines). Students will make more sense out of the technology by using the science they have just learned. Thus, students will grasp a deeper meaning of the science and the technology. More complex technologies can be introduced at this time.

Finally, the arrow in Figure 5.2 ends in the domain of society. Here students often address the original key question or social problem and then make a decision (Aikenhead, 1985a, 1991a). Students make thoughtful decisions informed by (1) an in-depth understanding of the underlying science, (2) a grasp of the relevant technology, and (3) an awareness of the guiding values. For instance, a class might decide what policy should be adopted by the local electricity board concerning high voltage power lines.

Figure 5.2 suggests a sequence that is useful for a lesson, a unit, or a textbook in STS science (Aikenhead, 1992c; Eijkelhof & Kortland, 1987). Variations in lessons or units can be achieved by starting a class in the technology domain (the black donut in Figure 5.2) with an interesting technology (for example, television satellite dishes). Alternatively, teachers can begin with an intriguing bit of science content, and move along the arrow from there.

Figure 5.2 does not indicate the time spent in the society, technology, and science domains. Perhaps 60% to 90% of the instruction time will be dedicated to the science domain. A teacher makes that decision, a characteristic of STS described by David Layton in Chapter 4.

In STS science, traditional science content is not watered down, but it embedded in a social-technological context. The choice of context can be made on the basis of (1) its meaningfulness to students, and (2) the science content that is logically generated by the context itself (on a need-to-know basis), as long as that science content is required by a particular science syllabus or is justifiable by the teacher. From the students' point of view, the science content seems to arise from, and is logically sequenced by, a real-life situation. This perspective lies in marked contrast to the traditional curriculum with its sequence determined by how an academic scientist would systematically conceptualize nature. STS science is student-oriented rather than scientist-oriented. The dichotomy between a scientist's view of nature and a student's view of the everyday world defines a fundamental difference between the traditional science curriculum and an STS science curriculum. In STS science teaching, traditional science content is certainly taught, but students learn this contentconstantly linking it with their everyday world (depicted in Figures 5.1 and 5.2).

SUMMARY

In Chapter 2, I argued that the character of scientific knowledge has changed over the past 100 years due to the socialization of science. Today we recognize science as having social aspects, both external and internal to science. These social aspects mediate the social construction of scientific knowledge.

STS science teaching conveys the image of socially constructed knowledge. Its student-oriented approach (Figure 5.1) emphasizes the basic facts, skills, and concepts of traditional science (John Ziman's "valid" science, Chapter 3), but does so by integrating that science content into social and technological contexts meaningful to students. The priority of goals for STS science teaching will vary in each country and community. A curriculum's STS content will alter accordingly. Nevertheless, I proposed a broad definition of STS content in order to help articulate the kinds of variations that exist in STS science teaching today.

The priority of STS goals may also affect the structure of an STS science course. Various structures are represented in Table 5.1. A recommended sequence in the design of STS science is suggested in Figure 5.2.

Good science-technology-society (STS) science education is relevant, challenging, realistic, and rigorous. STS science teaching aims to prepare future scientists, engineers, and citizens alike to participate in a society increasingly shaped by research and development involving science and technology.

Chapter 6

Teacher-Assisted STS Learning

Bill Williams

In the winter of 1983–1984, Rustum Roy and Paul Bell of Penn State University were awarded a National Science Foundation grant to support a project called Science through STS (S-STS). STS teaching was already happening in the high schools, some of it quite local, some of it state supported, and some in loosely coordinated groupings (for example, Irma Jarcho and her colleagues in New York producing the *Teachers' Clearinghouse Newsletter;* Roeder, 1993). The S-STS Project proposed to inquire into the extent of such STS teaching and to pilot a scheme for a nationwide STS K–12 organization in the United States.

This chapter looks at STS education as a grass-roots movement of classroom teachers committed to STS science. I begin with a brief history of the *S-STS Project,* an infrastructure that supported these teachers. This history reveals an essential ingredient to the success of their instruction. The chapter ends with two examples of typical teachers who illustrate this essential ingredient.

STS: A BRIEF HISTORY

Early in 1984, the newly formed Advisory Committee to the S-STS Project met. It comprised leading high school teachers with STS experience, and university, industry, and state scientists, technologists, and educationalists. It set up task forces to:

1. Produce a description of what STS comprises (a map of STS)
2. Describe the characteristics of good STS materials
3. Produce a model for a 1-year course in STS
4. Produce a catalog of available materials
5. Produce model teaching and learning texts.

The Advisory Committee also delegated various tasks to the headquarters staff: the two Principal Investigators, myself as consultant, a secretary, and whatever help we could obtain from Penn State's already overloaded STS faculty. These tasks included:

- setting up and distributing a quarterly newsletter, the *S-STS Reporter;*
- advertising its availability in teachers' journals, and thereby creating an address list of interested persons;
- investigating the possibility of holding a national conference.

It very quickly became apparent that interest and involvement in STS teaching was widespread across the nation. Our mailing list grew rapidly and soon exceeded 3000. We discovered that we had a lot of experience to draw on, even within our restricted pilot areas, at both the school and college levels. Within a year:

- the first three task forces had finished their work, and their findings were published in the *Reporter* (the catalogue of available materials and the model teaching and learning texts took longer to produce but followed in due course);
- accounts of various STS teaching, both in the United States and abroad, had been published;
- the *Reporter* had reviewed STS publications, measuring them against the task force's criteria;
- and arrangements had been made for the first Technological Literacy Conference (TLC–1) to be held in Baltimore in Feburary 1986.

The organization of TLC–1 was assigned to Leonard Waks, a new addition to the headquarters group. Its purpose was to bring together all those interested or involved in STS teaching (K–12 teachers; college teachers; local, state, and federal administrators; industrialists; journalists) so that there could be an extensive exchange of ideas and experiences.

To move from the first twinkle in someone's eye to the adoption of a science technology and society approach throughout a school system requires many components spread over several years. I think that the process falls into three stages:

Stage 1: Small groups of pioneers, seized with the conviction that a new approach is essential, invent, test, improve or reject, and try again, STS approaches. Teachers go through this cycle several times to discover whether their various inspirations are viable and, if so, how to make them work in the classroom.

Stage 2: Setting up a support system: the production of materials; establishing a support group; publishing a newsletter; holding local, regional, national meetings (conferences, summer schools, workshops); forming a professional association; and establishing respectability. And, when a substantial part of Stage 2 is in place, starting Stage 3.

Stage 3: Setting up a system to diffuse the STS approach into every school in the country.

K–12 teachers from across the nation, those with the twinkle in their eyes, were influential in every part of the S-STS Project, which was intended to deal with my Stage 2. And as the successive annual TLCs took place, evidence of their commitment and inspiration emerged more and more. But, strange to say, it was in Stage 3 that the strength and variety of local efforts became most evident.

Before the *S-STS Project* (Roy, 1988) was completed (it lasted 4 years), agreement had been given to a follow-up project, the *National STS Network* (Roy & Barchi, 1990). Barbara Barchi and Rustum Roy were the principal investigators, the base was again Penn State. The Network was to carry out the task that I have called Stage 3; that is, to extend the pilot to the remainder of the United States. To do this we set up a branching network and communications system. We divided the United States into nine regions, and engaged an associate in each to act as the base person for the regional activities and to organize a 5-day workshop in their region attended by 35–50 teachers or trainers. (A workshop in Puerto Rico was conducted on rather different lines.) We also continued to produce a newsletter, updated and extended the catalogue of materials (Cheek, 1992c), established a cost-free long-distance telephone number (an "800" number) for teachers to call for advice, and set up the basis of a computer communications network.

In the regional workshops we set out to trigger, to help, not to instruct. Workshop members were encouraged to develop their own thinking and their own materials and teaching aids. They were each given a large file (largely empty) divided into several sections: basic principles, materials, ethics and values, and the like. We told them, "Use this to build up your own collection of resources." We, the central group, provided: (1) guidelines (largely the products of the S-STS Project); (2) some model teaching and learning units (our own four, plus examples from the United States, Canada, the United Kingdom, and so on that the reader may find in the appendices to the model one-year course published in the *S-STS Reporter;* (Science Through STS Project, 1986); and (3) help, encouragement, and advice.

It is important in all teacher training, but particularly so in STS, to move away from dependence toward self-sufficiency. STS teaching and learning is essentially local, but not, of course, exclusively local. We stressed the need to start with the personal, move on to the local, and then to the national and the global. STS teaching and learning is also topical, student-centered, involving personal value judgements, not universal, change-resistant, value-free, top-down, passing-on-the-tablets. And as such the teacher must be an enabler as well as self-reliant to a considerable extent to be able to take advantage of topical issues as they occur. The STS teacher addresses such issues immediately, not postponing them to fit in with the syllabus.

In the New York regional workshop, for instance, global warming was the topic of the day and we had "classroom" discussions based on current media material, held a simulated public enquiry, and trawled the daily press for appropriate humorous items such as cartoons. The STS approach had to infuse the regional workshops themselves. In other words, start as you mean to go on—STS intends that student-centered learning should encourage inquiry, self-reliance, and the ability to form and defend one's opinion based on sound information and understanding, therefore, teacher training must use this model. What we were trying to produce was "ownership" (Hart & Robottom, 1990).

THE PRINCIPLE OF OWNERSHIP

I cannot overemphasize this principle of ownership. Teachers have much more interest in, commitment to, and enthusiasm for, something that they have helped to create than something imposed on them. To use the word *interest* in a different sense: they have a vested interest in making it work. This same principle applies to the students. Their involvement in activities and projects (particulary if, with a little judicious guidance, they are encouraged to choose their own projects) gives them a sense of ownership and produces a corresponding change of attitude.

Harold Hungerford, Trudi Volk, and John Ramsey, of Southern Illinois University, argued this most powerfully in their regional workshop at Carbondale. The application of this principle was demonstrated very effectively by workshop members Jim Porfilio and Jim Grover of Daniel Webster Middle School, Milwaukee. They and their colleagues go to great pains to give their students ownership—ownership of the curricula and indeed of much of the school's activities. The students and their parents are given a brochure listing the school's facilities, outlook, and programs, stressing the student-centered approach. Among the programs is the *Vital Issues Program* (VIP), their version of STS. And each student in that program carries a card bearing the logo VIP in 64-point type and the text "I am a member of the *Vital Issues Program* at the Daniel Webster Middle School. The bearer of this card is entitled to the rights, responsibilities and privileges of a very important person (V.I.P.)." And carrying that permit and under that banner they engage in local STS projects, social and environmental. I was particularly impressed by a program they produced for a radio broadcast on drug abuse.

For the STS training to spread outward from these workshops and to endure, it would not be effective to dispense recipes. The workshops had to set in motion a process that would help every teacher to construct his or her own program, to be able to adapt to changing situations, and to fashion her or his own tools. The workshops also had to encourage teachers to help each other. An important part of this is teacher-to-teacher communication; for example, local and state meetings, computer networks, and local newsletters.

To promote the diffusion process—the exponential growth without which we would never get to all those teachers—time was set aside for the group from each state to discuss their particular circumstances (urban, rural, industrial, agricultural, rich, poor, etc.) and to draw up a plan to extend the branching network to reach all the appropriate teachers in their state. Plans included training their own trainers, running their own workshops, writing their own materials, and making presentations at teachers' meetings.

There is some fear that STS founded on independent teacher-based curricula and materials is insecure and will not survive. For an innovation to persist, so the argument goes, there must be approval from some higher authority, an established curriculum, and fairly widespread—statewide or nationwide—agreement. Only then will the textbook publishers and examining boards cooperate. To be without their sanction is the kiss of death. But to kowtow to this "top-down" innovation model (Hart & Robottom, 1990) would be to revert in essence to the old established and, in my view, unsatisfactory pattern of science teaching, taking away from both teacher and student those very things we want to develop: critical thinking, decision making, value judgements, and so on. It seems very strange to me—and wrong—that we allow textbook publishers to dominate education and take away independence of mind from those actively involved. I have found the hundreds of STS teachers with whom I have had dealings, most in the United States and the United Kingdom but also in several other countries, to be independent, self-reliant, committed critical thinkers.

OWNERSHIP IN ACTION

All this is fine and dandy. But does it work? Can teachers develop their own materials and curricula satisfactorily or is that expecting too much?

The research that Margaret Wilsman (1991) and her team have done in Wisconsin shows very clearly that teachers have difficulty in using STS materials and resources—very fine materials and resources—supplied to them. Some experience more difficulty than others. The level of success depended on a number of factors, broadly speaking: teachers' prior experience with STS, their attitude, the extent of cooperation from and communication with their peers, and the degree to which their instruction focused on students' construction of concepts—"mediated instruction" On this last point, the researchers discovered that the most successful teacher taught in a way that corresponded to "category 5 STS science" described in Table 5.1 of Chapter 5, whereas the least successful teacher's instruction corresponded to a "category 1" viewpoint on STS (Wilsman, 1991).

These results are not altogether surprising. Any major change in teaching science will inevitably be difficult, and some will take to it better than others, as illustrated in Peter Fensham and Debora Corrigan's Chapter 18, in which they describe Australian chemistry teachers' adoption of an STS curriculum. Such a change will not be put in place overnight. It will require considerable preservice and inservice training. If we had a satisfactory science education system operational—satisfactory, not perfect—then we could be excused for saying "leave well enough alone." But we don't. Nobody reading Part I of this book could maintain that the present situation is satisfactory, largely neglecting as it does most of the population and not doing that well even for the more successful minority. The many failed attempts since World War II to change the current approach underline the need for some more radical change. STS holds out a hope of remedying this situation and is therefore worth pursuing, despite requiring major changes in knowledge base, teaching skills, and attitudes by science teachers.

So the question then becomes: Do we have any evidence that teachers can, separately and cooperatively, produce and handle STS materials and curricula to an acceptable standard? And, in the long run, we would also have to ask: Do the results, measured by better understanding, better skills, more enthusiasm, longer retention, and more scientific technological literacy throughout life, warrant the change to STS? I am not going to tackle that second question here. Glen Aikenhead begins the task in Chapter 16.

During the past 9 years, I have seen enough evidence in visits to schools, teacher conferences, printed material, and papers at the TLCs (over 50 K–12 teachers presented at TLC–8 in January 1993), to convince me that the answer to the first of those two questions is most definitely "Yes." My problem in presenting that evidence here is twofold: (1) I have collected enough teacher-produced materials to fill this chapter ten times over; and (2) I shall feel guilty about all those admirable products that I must inevitably leave out when providing a selection. For instance, two of our model

STS units were written by teachers. Jon Harkness and his colleagues at Wausau High, Wisconsin, produced a whole STS-based curriculum, which has been adopted and adapted throughout the state. I have seen teaching modules produced elsewhere in Wisconsin, Illinois, Oregon, California, Texas, New York, Kentucky, New Hampshire, Michigan, Iowa, Hawaii . . . need I go on? I have no doubt that there are many more that have not come my way. And then there are the excellent materials (1) from the Department of Defense (DoD) schools that the late Kent Rossier and his team wrote, tried, and tested, and (2) from many other places in the world where teachers have developed their own STS materials (many represented in this book), for example, The Netherlands, Britain, Canada, India, Germany, and Australia. It is an embarrassment of riches!

I single out two programs, written by the same two teachers: Jane Cappiello, Bethlehem Middle School, Delmar, New York, and Joanne Gallagher, Tamarac Middle School, Troy, New York. In selecting these I am not claiming that they are the best, whatever that may mean. They are a sample—not exactly a random sample—to illustrate why I believe so emphatically that teachers *can* produce and handle STS materials and curricula to an acceptable standard. *Science and Sports* is an eighth-grade course, lasting 10 weeks, at 3 hours class time per week. It has four parts: (1) "Society and Sports," dealing with the economic, political, social, educational, and careers aspects; (2) "Physics and Sports," the physical principles that underlie sports activities (motion, forces, energy, levers, and balance); (3) "Biochemistry and Sports," dealing with the chemistry and physiology of the body, including discussion of drugs; and (4) "Technology and Sports," tracks, gyms, stadia, surfaces, equipment, training, and injury rehabilitation. The STS character of all this is obvious. And equally obvious are the interesting and motivating nature of the subject and its potential for introducing the science and technology content in an attractive context. Each section contains activities such as writing projects and readings, and is lightened by experiments and humorous illustrations. Throughout there is a high "want to know" content: sports scholarships, nutrition, fashion, sports technique, career opportunities, the effects of steroids, protection against injury, and the features of good equipment.

The second module, from the same developers, shares many of the same characteristics, albeit in a very different setting. *Water,* adapted from a unified science supplement of the same title, has three parts: "Properties," "Sources," and "Uses." The first involves composition, surface tension, heat capacity, solvency, and density. In the second, students study where water is to be found and where it is not, the water cycle, surface water, and ground water. And in the third (the most extensive), students study the various common water uses, water management, and water pollution. In all three parts, there are student activities ranging from modeling a drainage basin, to water purity, to the amount of water wasted when brushing your teeth. The readings supplied are good, some of them fascinating (for example, the article "The Catalyst" from *Chemistry,* a fantasy about what might happen if ice were suddenly to become denser than liquid water, the normal density change for almost all substances). There is also a good bibliography. The module is not only opening up an STS approach, it is teaching traditional science content and improving study skills.

Science and Sport and *Water* are only two programs out of at least 100 I could have selected. Can teachers produce and handle STS materials and curricula? The evidence is compelling!

No wonder the *S-STS Project* found widespread interest and involvement when the Project began in 1983. Since then, support systems and diffusion systems have escalated the commitment to STS teaching by classroom teachers. This increase in commitment continues wherever teachers take ownership of their STS teaching.

Chapter 7

STS Teaching in Britain

Andrew Hunt

Mention STS to a group of British science teachers and you are likely to mystify your audience. The term does not crop up in everyday staffroom conversations. Other acronyms such as SATIS, SISCON, or SCISP are much more likely to be greeted with nods of recognition.

Enthusiasts have developed and nurtured approaches to STS for 20 years or more. Now they are managing to infiltrate their ideas and teaching methods into parts of the mainstream science curriculum. Even so, progress is slow, halting, and seldom based on a clear rationale. Nationally we share in John Ziman's confusion and have, in various ways, tried all the approaches he describes in Chapter 3. This chapter traces these developments and the impact of STS schemes on teaching science in British classrooms.

EARLY SEEDS

Nuffield is the name we associate in Britain with the beginning of the era of curriculum development projects in the 1960s. Best known are the specialized, prestigious, Nuffield schemes featuring highly academic physics, chemistry, and biology. Many of those in the working group that devised our new National Curriculum of today began their teaching careers as enthusiastic supporters of these schemes. This Nuffield influence lives on.

In Nuffield Physics (1967), applied science was a diversion used to provide occasional motivation to enliven the course. Teachers generally expected their pupils to be thrilled with the idea of scientific experiments and knowledge. Less well known was another 1960s Nuffield program that was much more important as a precursor to some of the STS strands now visible in school science. The Nuffield Foundation funded its Secondary Science Project in response to the need to develop a science curriculum for pupils between the ages of 13 and 16 of average and below-average ability. In the development work and field trials, the development team looked for science topics and themes that had immediate significance for the pupils in terms of its intrinsic interest and, at the same time, were concerned with the real adult world (Nuffield Secondary Science, 1971). Anything that failed the significance test was deleted.

This was only one of several moves to make science more relevant. The teaching strategy was also built around a series of engaging practical activities linked to eight themes covering everyday materials, human biology, environmental topics, and the harnessing of energy resources.

As John Ziman points out (Chapter 3), relevance is a necessary but not sufficient condition for STS. Nuffield Secondary Science was not an STS scheme, but it stimulated much creative work leading to new definitions of "science for all." From the STS point of view, however, it established an unfortunate association of relevant science with less academic pupils. The two Nuffield traditions illustrate only too well Young's (1971) proposition that in those days only two types of curriculum change seemed possible: (1) those in which existing academic curricula are modified but there is no change in the existing evaluation of knowledge, and (2) innovations that disregard the social evaluations implicit in academic curricula, but are restricted to less able pupils.

GREEN SHOOTS

In the early 1970s, there was a brave attempt to win converts to a transdisciplinary approach to mainstream science education. This was the *Schools Council Integrated Science Project,* SCISP (1973), the first scheme to make a big feature of the social implications and technological applications in a science program for the younger students (14 to 16 years old). The authors extended their coverage to include aspects of earth sciences, psychology, and social sciences. They argued that problems concerning the effects of pollution, the availability of fuels and food, transport and disease, make nonsense of discipline boundaries because these problems involve the whole science spectrum, from the social sciences through to physics (Hall, 1973). A small number of enthusiastic teachers adopted SCISP. The project helped the cause of STS without winning a large following for integrated science courses.

In the mid-1970s, science teachers recognized a fresh challenge. Despite the new Nuffield courses, young people were still turning away from science, which many associated with the bomb, pollution, and genetic manipulation. Some of the teachers who had been active in the Nuffield schemes surmised that they had enjoyed themselves too much by investigating the intellectual niceties of theoretical science, when they should have been more concerned with the interplay between science and society

(Lewis, 1978). At the time, many science teachers felt ill-equipped to present a broader view.

In response to this need, the Association for Science Education set up the Science in Society project in 1976. Looking for a home for the new scheme in the curriculum, the project team identified General Studies lessons for 16- to 18-year-old students as a suitable niche. This allowed freedom for innovation without threatening the recognized routes to advanced qualifications (Association for Science Education, 1981).

The publication series *Science in Society* (Lewis, 1981) was taken up by many private and selective schools. In doing so, these schools played an important part in winning recognition for the place of STS in the science education for all. It attracted support from industrialists who hoped to see schemes of this kind help establish a broader curriculum that would be a preparation for future engineers and managers. They hoped that a broader understanding of science and technology would encourage more young people to contemplate careers in industry. In this sense the scheme had a "vocational" appeal to its sponsors, in the sense described by John Ziman in Chapter 3. The principal aims of *Science in Society* were to help students to (1) understand the nature and limitations of scientific knowledge, (2) appreciate that the use of scientific knowledge can be both beneficial and detrimental to society and the environment, (3) appreciate that the earth's resources are finite, and (4) develop the ability to make reasoned decisions that take account of relevant constraints and to recognize that moral considerations are involved in making decisions.

Critics of the program felt that it was too closely identified with the views and values of industry, corporations, and professional institutions. The authors of the course materials consulted many expert advisers but generally did not seek to involve those holding opinions that challenged the established order.

In 1978, a parallel STS development began to colonize the same general studies niche (16- to 18-year-olds). It grew from a project to promote STS teaching in universities and polytechnics. In higher education, SISCON (Science In a Social CON-text) linked science with economics in both developed and developing countries, drew on the new studies of the philosophy and sociology of science, and studied the connection between science and warfare.

The *SISCON-in-Schools* team did not hesitate to raise topics with wide political implications. They hoped that, by confronting controversial issues, high school students would learn to be mature and active participants in their technological society (Solomon, 1985). In the course materials, students explored not only the impact of science and technology on society, but also ways in which the cultural context has a strong influence on the direction in which science develops (Addinell & Solomon, 1983).

The *SISCON-in-Schools* readers (Solomon, 1983) were full of lively stories of discovery and invention presented in a way that avoided the academic dryness that John Ziman warns against in his discussion of the historical approach (Chapter 3). Case studies from the history of science and technology helped students learn about (1) the nature of scientific explanations and the ways in which theories change, (2) examples of technologies arising from the needs of society and the impact of these technologies on ways of living, and (3) the interaction between industry and all aspects of society including war, medicine, and education.

The numbers involved in *Science in Society* (Lewis, 1981) and *SISCON-in-Schools* (Solomon, 1983) were small. The schemes were important, however, because they opened the eyes of science teachers to the possibilities and rewards of STS. Equally important, perhaps, was the effect they had on teaching methods. Science teachers found themselves exploring new methods including group discussion, simulations, role-play, and games.

TAKING ROOT

A national policy for science education began to emerge with the arrival of a new examination system for 16-year-olds in 1986. Mandatory criteria for the GCSE (General Certificate of Secondary Education) required all examinations to give at least 15% credit for the applied, social, and technological aspects of science, despite the fact that these were seldom treated adequately in existing textbooks.

Knowing that the GCSE was on the way, the Association for Science Education took the initiative to launch a new project. The aim was to develop resources to support teachers who were obliged to introduce aspects of STS into their courses for the first time. The outcome was the *Science and Technology in Society* (SATIS) publications (Hunt, 1988).

The SATIS project had a central writing team but invited teachers from all over the country to send in contributions—not just ideas for interesting topics but also ideas for effective classroom strategies. The team decided to produce short, easy-to-use units that teachers could easily infuse into their mainstream science courses. In selecting ideas for trials and publication, the editors preferred units that (1) showed strong links with examination topics, (2) featured a modern application of science or a topical issue, (3) were developed in collaboration with an expert adviser, and (4) provided engaging activities for the students. SATIS places great emphasis on the "relevance" approach (Chapter 3), by contributing materials that are relevant to: the future citizen, the future worker, the student's own life, and the wider world (Holman, 1986).

Teachers in Britain (and to some extent North America) quickly adopted the SATIS materials because the materials met an urgent teaching need. At first the project did not have a coherent STS rationale (thus, its status as a "category 2" project in the STS science scheme found in Chapter 5). Some of the early units about the applications of science and technology barely qualify for the STS label. Even so, the rapid uptake of its ideas meant that SATIS soon had a big influence on curriculum development and textbook writing. It also legitimized teaching approaches that were previously rare in science classrooms. Now it is no surprise to find a class engaged in debate as students simulate "The Limestone Inquiry," or to see small groups of students discussing the implications of scientific data such as the results of "Tristan da Cunha Dental Surveys," or the problems of "Drinking Alcohol." Thanks to SATIS, many more science teachers expect to spend some time encouraging students to explore environmental threats to the quality of life, to develop decision-making skills by studying the economic and industrial aspects of technology, and to formulate opinions about controversial issues. In these ways, SATIS was an important advance for STS in Britain.

SATIS gained national recognition at a time when professional scientists and industrialists were promoting the need to increase the general "public understanding of science" (described by Joan Solomon in Chapter 10 and Arthur Lucas in Chapter 11). This made it relatively easy to win more support and substantial funding for a second project, this time SATIS 16–19 for older students (Hunt, 1988; Hunt & Lenton, 1992). The members of the new team faced a double task. One objective was to extend the SATIS model by providing infusion materials for more advanced courses. Another was to support general education for specialist and nonspecialist students. For both objectives an STS rationale was needed. Consequently, Joan Solomon (1992b, 1992d, 1992e) wrote three short readers that became the framework for the project. SATIS 16–19 evolved into a "category 3" STS project, according to Table 5.1 (Chapter 5).

All infusion units accepted for trials and publication in SATIS 16–19 had to explore, in one way or another, topics or issues featured in one or more of the readers. For example, to develop the ideas in *What is Science?* (Solomon, 1992d), the project team produced units such as: "The Retrial of Galileo," "Thinking and Learning about Science," and "Patterns in the Sky." To provide activities that help answer the question *What is Technology?* (Solomon, 1992e), SATIS 16–19 published: "Kidney Transplants," "Appropriate Technology," "Energy from the Waves," and a historical unit, "William Perkin—Founder of the Synthetic Dyestuffs Industry." To foster decision-making skills and an awareness of the social context of technical issues (*How Does Society Decide?* Solomon, 1992b), SATIS 16–19 devised many units, including: "Animal Rights and Animal Wrongs," "How Shall We

Live?" "Risk," "After Chernobyl," and "Medicines in Developing Countries." Thus, many units in the SATIS and SATIS 16–19 publications deal with issues related to the cluster of problems that John Ziman (Chapter 3) refers to as components of a world *problematique*.

GROWING INTO A SEEDLING

In the late 1980s, the British government took much more interest in education policy. A reform act in 1988 began the writing of a National Curriculum covering the 5 to 16 age range. At first sight, the science curriculum seems to offer much to teachers wishing to place emphasis on STS. The general introduction to the program of study for 14 to 16 year olds insists that students should be given opportunities to (1) develop their knowledge and understanding of how scientific ideas change through time and how their nature, and the use to which they are put, is affected by the social, moral, spiritual, and cultural contexts in which they are developed; (2) study how science is applied in a variety of contexts; (3) use their science knowledge and skills to make decisions and judgments concerning health and safety; and (4) consider the effects of scientific and technological developments on individuals, communities, and environments.

This is clearly an STS agenda and it is supported in the detailed programs of study by the requirement that 14- to 16-year-old students should carry out investigations in which they explore the nature of scientific evidence and proof, study examples of scientific controversies, and study how scientific ideas change. Students are also expected to explore a wide range of technologies and issues including: the social, economic, and ethical aspects of cloning and selective breeding; current concerns about human activity leading to pollution and its effects on the environment; the beneficial and harmful effects of radiation; the social, economic, environmental, and safety factors associated with the manufacture of materials; and the longer term implications (including the greenhouse effect) of the worldwide patterns of distribution and use of energy resources. Overall, the program requires students to weigh evidence and form balanced judgments about some of the major environmental issues facing society.

My selection of National Curriculum objectives mentioned above might suggest that the new National Curriculum has adopted an STS approach. This suggestion underestimates other pressures that dictate the way in which the curriculum is taught in schools. The dominant influence is the system of national examinations at ages 14 and 16. So far, these tests have concentrated on the more traditional and academic concepts. The assessment strategies are not well suited to crediting students for their exploration of STS issues.

One interpretation of the National Curriculum for 14- to 16-year-olds comes in the form of 13 publications from the Nuffield Modular Sciences (1992–1993) project, which weave together many of the development strands referred to in this chapter. The publications draw on the original Nuffield science courses for practical approaches to higher level concepts, but set the science in many of the contexts identified as relevant by Nuffield Secondary Science.

The National Curriculum tends to emphasize the separation of the sciences, but some of these new Nuffield modules offer a degree of integration and, like SCISP, include aspects of earth sciences, astronomy, and (where the curriculum permits) some mention of behavioral sciences. SATIS-style activities help students take part in discussions, simulations, and debates, while case studies and stories enliven and humanize the science in the spirit of *SISCON-in-Schools*. The writers of Nuffield Modular Sciences have infused STS aspects in a systematic way, but an STS rationale contributes to the framewwork of only a limited number of the 13 modules. The Nuffield Modular Sciences project caters to the wide range of aptitudes, interests, and thinking styles of the students in a typical secondary school. The aim is to provide a single balanced science program for all, both future citizens (an STS emphasis) and prospective scientists and technologists (a more traditional Nuffield emphasis).

CONCLUSION

The growth of STS teaching in Britain has evolved through a succession of projects, each with its own set of values and features passed on to the ensuing project: from the seeds of a science-for-all philosophy in Nuffield science, to the relevance of interdisciplinary SCISP with its expanded student evaluation philosophy, to the vocation appeal of *Science and Society* with its decision-making teaching methodology, and then to *SISCON-in-Schools'* critical analysis of science and culture with an even greater repertoire of teaching methods. A wider implementation of these values and features is anticipated for SATIS and the Nuffield Modular Sciences, as STS teaching continues to evolve in Britain.

Chapter 8

Developing a Model for Issue-Oriented Science

Herbert Thier and Barbara Nagle

The desire and demand for change in science education is growing world-wide, as evidenced by Project 2000+ (UNESCO, 1993) and similar efforts by national and international bodies. Many prominent individuals and groups in the United States (for example, the National Science Foundation, 1983) are calling for a major science curriculum reform to prepare students to understand the roles of science and technology in their personal lives and in the growth and development of the nation. Accordingly, science education must present concepts in a context that is centered around the needs and interests of the learner, and must provide an experience-based understanding of the impact and importance of modern science and technology.

Previous chapters in this book have described such goals for STS science programs and have suggested a range of instructional approaches. In this chapter, we will look at the approach to STS called "issue-oriented science" taken by the *Science Education for Public Understanding Program,* SEPUP (SEPUP Annual Report, 1992). The chapter examines the unique features of the SEPUP approach and discusses how the program relates to the spectrum of STS science described by Glen Aikenhead in Chapter 5 and how it embraces the interdisciplnary approach characterized by John Ziman in Chapter 3.

SEPUP arises from the successes of an earlier STS curriculum, the *Chemical Education for Public Understanding Program,* CEPUP (Thier & Hill, 1988). Both the content and instructional approach of CEPUP were designed to improve scientific and technological literacy, and to provide a more effective and motivating science curriculum for the needs and interests of the widely diverse population of students in the United States. SEPUP's approach to achieving these needs is based on the concept that the science teacher is the key to promoting effective long-term change. In the typical middle school or junior high school, the science teacher is the only one who has a background in science. As educational budgets shrink nationwide, many school districts in the United States have no other science leadership. Therefore the responsibility falls on the science teacher to provide the leadership. Only then will there be the school involvement necessary to accomplish scientific and technological literacy.

The name change from "chemical" to "science" education (CEPUP to SEPUP) reflects the expansion of topics in the new phase of our work. This new phase is, at the time of writing, in the process of being developed and has a completion date of 1996. When finished, SEPUP will have published two year-long, issue-oriented science courses for the early secondary school (grades 7–10). In this chapter, CEPUP refers to the first phase of the project (now completed), including 12 CEPUP modules, and SEPUP refers to the overall project (to be completed in 1996), including the two new courses, materials, and development efforts.

ISSUE-ORIENTED SCIENCE: THE *SEPUP* APPROACH

SEPUP shares several characteristics with many STS programs: a concern with scientific literacy, movement towards a science curriculum that reflects societal values and is relevant to student lives, and the desire for a curriculum to engage and challenge all students. SEPUP starts with the assumption that the current "saber-tooth" curriculum (described in Chapter 2) is totally inadequate, not only in its division of science into three artificially defined disciplines, but also in its conceptual emphasis. In SEPUP, the issues and the science are inextricably connected.

We use the term "issue-oriented science" to describe SEPUP because the term "STS" has been used to describe a wide variety of programs. These range from a controlled infusion of material into a traditional science course to advocacy programs for one cause or another that barely resemble science. Issue-oriented science in SEPUP is not an add-on that explores societal concerns only on Fridays, nor does it aim to foster student activism for a particular environmental cause (thus avoiding problems that arise in citizen-action courses, problems identified by David Layton in Chapter 4). Issue-oriented science has as its goal the development of an understanding of the science and the problem-solving processes related to social issues, without taking an advocacy position.

The science in an issue-oriented science classroom is not read-the-textbook-and-answer-the-questions science. It involves activities and investigations, discussions and debates, themes, and questions. Students in issue-oriented classrooms learn to understand scientific evidence and its limitations, assess risks and benefits, ask questions, and make decisions based on evidence rather than on pure emotion.

Teachers in issue-oriented classrooms create situations where understanding can grow, where issues can be explored, and where students can interact. The teacher models tentativeness, objectivity, and nonadvocacy, and is willing to say, "I don't know, let's find out" and "Science cannot answer that question, it's a public policy issue." Whether students are discussing the choice between paper or plastic bags in the supermarket, formulating a waste management plan for their community, or writing an environmental impact report for a factory on an imaginary island, SEPUP does not tell students what decision to make. The SEPUP approach facilitates understanding of the available evidence so that students can come to their own conclusions about the issues. By maintaining a nonadvocacy position, SEPUP avoids some of the pitfalls of the problem-oriented approach described in Chapter 3. This kind of science education helps students develop the knowledge and judgment necessary to make effective, evidence-based decisions regarding both their own health and welfare and the variety of science-based public issues they will encounter as participating members of a democratic society.

Although the two SEPUP courses will most closely resemble categories 5 and 6 of Glen Aikenhead's spectrum of STS science (Chapter 5), the materials in CEPUP modules (now completed) are different from each of these categories in significant ways. For example, CEPUP modules fit Glen Aikenhead's category 5 (science through STS content), in that the issues in a CEPUP module organize the science content that is necessarily multidisciplinary in nature. CEPUP modules resemble category 6 (science along with STS content), in that students are assessed equally on the issues and the pure science content. However, in SEPUP, although the issue will provide the reason for delving into a particular area of science content, the science content will not be viewed as enriching the STS content, but as providing absolutely essential evidence, necessary for an understanding of the issue and possible decisions. The central role of evidence-based decision making results in a program in which key science concepts and processes are studied in depth. As we move from CEPUP modules to SEPUP courses, the development of key science concepts (such as the particle nature of matter, chemical concentration, and the ways in which substances interact with living organisms) assumes even more importance in providing students with the background to understand the issues. In the beginning of the first SEPUP course, students will build a "scientific tool box" to develop an operational understanding of concepts such as acids and bases, risk, and qualitative and quantitative analysis, so that they will be able to use these ideas to investigate science-based societal issues of interest and concern to them in the rest of the course.

The SEPUP Courses

SEPUP is developing two year-long courses for the early secondary school years (14- to 16-year-old students) with a 1996 completion date. The transition from CEPUP to SEPUP evolved as a result of experience with the broad implications and interrelationships of the issues raised in CEPUP modules. The study of chemical issues and their implications for society resulted in activities that address the traditional science disciplines of earth science, chemistry, physics, and biology in an integrated fashion. For example, the examination of soil and groundwater issues (such as chemical contamination and hazardous waste disposal) resulted in activities that relate to earth science. Comparison of the physical properties of different plastics resulted in activities involving physics concepts such as heat transfer and tensile strength. While studying a two-step reaction for a proposed chemical factory, students learn such chemistry topics as the factors controlling reaction rate, the evidence for chemical change, and chemical equations. Examination of risk management, chemical residues in food, the role of microorganisms in food poisoning, and toxicity testing resulted in activities that involve biological science concepts. These are only a few of the examples from CEPUP modules that illustrate the power of issue-oriented science to connect and integrate the traditional scientific disciplines.

Both proposed SEPUP courses will focus on issues involving the impact of personal and public decisions on the world. The issues might be quite personal for students, such as whether to use paper or plastic bags ("Plastics in Our Lives" module) or whether to buy organic produce. They may be community-based, such as deciding whether to build a hazardous waste incinerator ("Waste Hierarchy" module), or even global, such as determining policies regarding the use of chlorofluorocarbons (CFCs) and their impact on the stratospheric ozone layer ("Refrigeration" module). One aim of the courses is for students to realize that these issues are complex, that there usually are no easy answers, and that therefore trade-offs need to be considered. The current CEPUP "Island Factory" module is a good example. After learning a great deal about chemical interactions and what is required to scale up an experiment into an industrial process, students deal with the trade-offs involved in bringing a new industry to an island that is known for its beauty as a tourist spot but needs additional jobs to support its population.

The two SEPUP courses could be the basis of a model for schools in any country to use to emphasize issue-oriented science in their own curriculum. When completed, either course will be able to stand alone, be used in sequence, or serve as a resource for those who wish to prepare a customized course to respond to local needs.

We think of the first SEPUP course as a foundation course for the middle school or junior high school years in the United States. It will focus on concrete and experience-based issues that affect either the students' personal lives or their local community. More than half the material for this course will be adapted from existing CEPUP modules. Additional materials will include activities that encourage students to interrelate their learning from several of the CEPUP modules, and to become more independent in their design, analysis, and written and oral reporting of laboratory experiences.

The second SEPUP course will stress more abstract issues, those that have global implications and involve difficult trade-offs. It will emphasize a global perspective by highlighting the concept of sustainable development and the need for risk management in developing scientific and technological solutions to societal problems. Sustainable development is defined by the United Nations-sponsored World Commission on Environment and Development (1987) as development that meets the needs of the present without compromising the ability of future generations to meet their own needs. This concept provides a bridge to a global view of societal issues. Sustainable development considers the viewpoints and interactions of various individuals, groups, and nations. Taking this point of view in an instructional program encourages students to evaluate personal and community decisions in the context of their impact on current and future society.

Topics under consideration for the second SEPUP course include comparing environmental health risks, biotechnology, global implications of energy sources, and global atmospheric changes. The overall goal of SEPUP is to contribute to "a new social contract between science and society [that] seeks a balance between, on the one hand, power and dominion over nature, including economic well being, and on the other hand, stewardship of the earth and quality of life" (Aikenhead, Chapter 2).

Goals of the SEPUP Courses

The goals for the SEPUP courses under development are described here. (Readers interested in reviewing the specific goals and objectives of the CEPUP modules are invited to write to the authors for a copy of "Issue Oriented Science Using CEPUP," which includes a synopsis of the content of each module.) The overall goals of SEPUP are:

1. Develop learner-centered instructional materials to make students more independent as thinkers and active participants in science. The materials should provide a model for secondary science reform that is flexible and responsive to local school needs and appropriate for students of all ethnic, cultural, and socioeconomic backgrounds, especially students who are at risk.
2. Enhance the role of teachers as facilitators of student learning and as educational leaders within their communities by having them share in the development, implementation, and assessment of issue-oriented science materials and programs.
3. Facilitate the development of competence and independent learning by encouraging students to use scientific information and risk analysis in personal decision making. Activities in each course should emphasize the critical analysis of experimental data and provide progressively more open-ended experiences for students as they develop the capability, through SEPUP, to learn independently.
4. Integrate science concepts and processes with relevant language, writing, mathematics, and social science activities in the school as a whole through subject integration, schoolwide projects, or both. The proposed student journals and other interdisciplinary activities in the courses, as well as continuing staff collaboration with individuals and professional associations in other disciplines, will contribute to the accomplishment of this goal.
5. Assess the effectiveness and impact of the SEPUP materials. Research will assess the effect of SEPUP on student and teacher attitudes, behaviors, and decision-making processes, as well as the impact of the program on student learning in a variety of classrooms serving widely diverse student populations. This research effort will include the development of models and materials appropriate for authentic assessment of issue-oriented, activity-based science, as called for by Champagne and Newell (1992).

More important than the products of curriculum development is the rationale used to develop the products and the impact of that rationale on the thinking and actions of the people involved in the project, from the core staff to the reflective user. Materials, and the instructional development projects that produce them, usually have relatively short lives. The long-term effects are the changes in people brought about as they internalize and restructure the overall goals and objectives of the project.

INSTRUCTIONAL DESIGN OF ISSUE-ORIENTED SCIENCE MATERIALS

The National Science Teachers Association, (NSTA) (1991) recommends that at the middle school level, a minimum of 80% of the science instruction time be spent on laboratory-related experiences, including prelab preparation, postlab discussion, and analysis. They also recommend that investigations should relate to current issues in science and technology.

Several studies cite the effectiveness of activity-based science programs (Bredderman, 1983; Shymansky, Hedges, & Woodworth, 1990) and science programs that allow students to relate content to their own experiences (Spector & Gibson, 1991). Experiences with such programs are especially advantageous to historically underrepresented students (Bredderman, 1982; Hager & Von Braun, 1989) and to girls, who report less experience than boys with the tools of technology and experimentation (Kahle, 1988). Despite these findings, science lessons in the United States usually include lecture, discussion, and reading from a textbook, while hands-on or other experience-based activities are infrequent. According to Yager (1983), 90% of students use a textbook 95% of the time in science classrooms.

The SEPUP courses under development, like the CEPUP modules on which they are based, will *not* include a textbook. This does not mean that the importance of reading, language development, and communication in science will be diminished. Instead, the courses will contribute to the development of the students' scientific literacy by focusing on experience-based inputs and using an innovative approach to presenting appropriate print and other materials. Each course will include the following materials:

- *A teacher's resource book.* The teachers' resource book will include extensive background materials, complete materials for carrying out the activities and leading discussions, and suggestions for extension activities and homework. Authentic assessment activities will be integrated into the teaching program.
- *A nonconsumable student book.* The student book will contain student readings and illustrations to enrich the program, complete instructions for carrying out laboratory experiments and other activities, and provocative questions for students to answer in their written reports of their laboratory activities.
- *A student journal.* The journal, provided by the student, will contain the student's notes and data tables from the laboratory activities, laboratory reports, and other written responses to assignments.
- *Projects and extension activities.* Activities involving students in actual research will be included in the courses. Additional open-ended

questions will be provided as a stimulus for those users who wish to perform extension activities. Students will be encouraged to develop their own projects.

- *Kits*. The program will continue to be kit-based. Kits will include all materials in convenient containers or dropping bottles, labware, and all other materials needed for activities, except for very common classroom items and perishable items, such as scissors and fresh food. Technological aids to instruction, such as videotapes and computer-based methods for including recorded evidence in the course, will be used as necessary and feasible. All such materials will be part of the kit.

SEPUP will continue to enhance the CEPUP approach of designing student materials that can be used by students with a wide range of skills and learning styles. For example, there is the 4–2–1 instructional model for laboratory activities (Thier, 1991) in which groups of four students share materials, pairs of students share a CEPUP tray and perform experiments together, and each student records the results of the activity and responds to questions. The cooperative learning encouraged by this model and the variety of activities in the program (including brainstorming, discussion, and role-playing) encourage students of varying learning styles to participate successfully in the program.

The expectation that these approaches to instructional design will be successful is reinforced by a major assessment study (Kelly, 1991) that shows a statistically significant positive impact from the regular use of the CEPUP modules on student attitudes toward science and achievement of science concepts and on teacher attitudes and actions. Results indicate that CEPUP significantly enhanced students' intentions to take more science classes in high school, students' perceptions that what they learn in science class is useful in everyday life, and students' expectations of the need for a knowledge of science in their life work. Results of achievement tests on module content indicate that a majority of students acquired an understanding of the wide variety of concepts emphasized in the program. Teachers using CEPUP modules reported increases in the following classroom activities: problem solving and discussion of environmental and community issues, including making decisions about these issues. The teachers also report increases in their own professional activities, such as attending science content inservices, participating in curriculum development efforts, and cooperating with science teachers in other schools.

SUMMARY

SEPUP draws its content and emphasis from the world students live in and from the science concepts, processes, and applications students need to know. We view SEPUP as science for citizenship in the 21st century, and we consider the concepts central to SEPUP more important than some of the concepts considered essential in the traditional curriculum. For example, some of the key concepts in SEPUP include: risk assessment and risk-benefit analysis, the collecting and evaluation of evidence and use of this evidence in decision making, the methods used to detect and measure the levels of chemicals in the environment, the principles of animal testing as a means to determine toxicity, and the use of life-cycle analysis to make choices about the best materials and processes to use to meet our desire for new products and services. Surely these concepts are important enough in modern life that teaching about them should take precedence over the "saber-tooth" curriculum (Chapter 2) that unfortunately is not extinct in many of our schools.

Chapter 9

Teaching *ChemCom*: Can We Use the Text Without Being Used By the Text?

William Carlsen, Gregory Kelly, and Christine Cunningham

This chapter reports on a research analysis of a precollege chemistry program *ChemCom: Chemistry in the Community* produced by the American Chemical Society, ACS (1988a, 1988b). As an STS curriculum, *ChemCom* is noteworthy in three ways:

1. Although the textbook was produced by a mainstream scientific organization with over 100 corporate members from chemical-related industries, it presents a view of science and society that is consistently multivocal, providing detail about positive and negative consequences of science and technology;
2. The textbook is pedagogically imaginative, devoting many pages to activities such as simulated public hearings;
3. *ChemCom* teaches chemistry on a need-to-know basis by focusing attention on important contemporary social issues, including water quality, food safety, waste disposal, and energy.

Of the approaches described by John Ziman in Chapter 3, *ChemCom* exemplifies "relevance" and "problematics." *ChemCom* is a "category 4" course in the scheme presented in Table 5.1 of Chapter 5.

The explicit goals of the curriculum are stated in the textbook's preface:

> This year-long course is designed primarily to help students:
> - Realize the important role that chemistry will play in their personal and professional lives
> - Use principles of chemistry to think intelligently about current issues they will encounter that involve science and technology
> - Develop a lifelong awareness of the potential and limitations of science and technology. (American Chemical Society [ACS], 1988a, p. ix).

Given the *ChemCom* project's sources of financial support (The National Science Foundation, ACS Corporation Associates, and the Petroleum Research Fund) and the project's agenda of preparing students for participation in democratic decision making, we wondered how science would be represented and whose interests would be served by the curriculum. We were concerned that the text might convey social and political messages in an almost subliminal way, and thus use a chemistry teacher as an ideological conduit for those hidden messages.

Can we use *ChemCom* without being used by the text? Yes we can, as long as we fine-tune three of our critical capabilities:

1. Our analysis of the underlying philosophy of the textbook, found in curriculum documents and the text itself;
2. Our understanding of the social nature of science, particularly the sociology of science, but also the philosophy and political economy of science;
3. Our detection of arguments and rhetorical strategies that express social messages about the interrelationships among science, technology, and society, and therefore about the sociocultural aspects of science.

These three capabilities frame the content of this chapter.

The chapter has two main purposes. First it provides guidance to researchers interested in developing their own critical capabilities. Secondly, the chapter describes some results from our investigation of *ChemCom*.

Our investigation took a full year to complete. The first author taught *ChemCom* in a local high school where he helped plan instruction with other teachers. We all participated in the critical analysis schemes. Our work unfolded in three phases: (1) a general curriculum analysis, (2) a general description of STS interrelationships, and (3) a sociocultural curriculum analysis, designed and tested on *ChemCom* but applicable to all STS curricula. Each phase is described, with emphasis given to phase 3.

PHASE 1: GENERAL CURRICULUM ANALYSIS

We scrutinized the *ChemCom* curriculum by applying Posner's (1992) analytic framework to many of *ChemCom's* supporting documents and reports (catalogued in Carlsen, Kelly, & Cunningham, 1992). We described the origin, purpose, content, assumptions, theory of learning, and implementation concerns that underlie the curriculum. Our analysis attended to the philosophy underlying the curriculum: its views of chemistry, science, students, teaching, and education. Close study of the textbook, its organization, and text-embedded activities, revealed both official goals and a hidden curriculum (Cunningham, Carlsen, & Kelly, 1992). Examples of each are outlined here.

The first unit in *ChemCom,* "Supplying Our Water Needs," investigates a mysterious fish kill in the fictional town of Riverwood. Chemistry information, analytical techniques, and water-use data are introduced throughout the unit in order to help students gather information about the cause of the kill. As the culminating unit activity, students hold a mock Town Council meeting to ascertain who is responsible for the fish kill and who should pay the costs associated with the emergency. Students are assigned to one of six special interest groups that will testify at the meeting. To assist students with their roles, the textbook provides a few paragraphs for each position, summarizing important facts or viewpoints that students might use in their arguments.

The data provided throughout the unit converge to identify the problem as "gas bubble disease" caused by the local hydroelectric power plant. The information leaves little room for debate. The power company holds sole responsibility for the fish kill, having unintentionally caused supersaturation of nitrogen and oxygen below the dam, in excess of legal limits. The facts provided by the simulation to "Power Company Officials" corroborate the "Scientists'" testimony. No fact challenges the scientists' findings or stimulates debate. Finally—and incidentally—the "County Sanitation Committee" reports an increase in another water contaminant, selenium. Although the contaminant provides some cause for concern, it is clearly not a contributor to the fish kill. Significantly, the source of the selenium is not agricultural runoff, wastewater discharge, or point-source pollution by any business or industry along the river. Its origin is natural: "Selenium-containing soil near Riverwood came from the debris of prehistoric volcanic eruptions" (American Chemical Society [ACS], 1988a, p. 74). To summarize, unambiguous and uncontested scientific data show that the fish were killed by a renewable energy resource. Chemists called to the scene discovered data that may ultimately protect the fish from another potential killer (prehistoric volcanic eruptions).

The text could have provided speakers with a wider range of information and perspectives. This might have encouraged a public hearing that questioned the scientific evidence or explored typical causes of fish kills. The purpose of this debate, however, is not to debate the science; it is to debate the social consequences that follow. Conspicuously, scientists are the only group given quantitative data. The "Chamber of Commerce," for example, is given no specifics about financial losses. Instead it is told that damages from the fish kill are "a substantial sum" and that "total financial losses due to the water emergency may be much higher than currently projected." What is likely to happen at the public hearing? Objective and dispassionate facts will be presented by scientists, while vague and emotional claims of damages will be presented by the townspeople. Meanwhile, agricultural and industrial chemistry escape the chapter unscathed, and the status of scientific facts as unimpeachably objective remains unchallenged. The unit concludes with comforting language:

> The Riverwood water mystery has finally been solved. It is ironic that the water "contaminants" responsible for the fish kill and for the town's understandable concern are just excessive amounts of oxygen gas and nitrogen gas dissolved in the water. Neither substance qualifies as toxic or dangerous—after all, we live immersed in an atmosphere of these gases. (American Chemical Society [ACS], 1988a, p. 74)

Instead of resolving the fish kill mystery, this passage raised many new questions for us. Who benefits from the "irony" written into this fictional mystery? Why is the word "contaminants" in quotation marks? And can one read a slightly patronizing tone in the juxtaposition of "the town's understandable concern" with the expression, "just excessive amounts of oxygen gas and nitrogen gas dissolved in water"? In the unit "Supplying Our Water Needs," does *ChemCom* reify the status of scientific facts, promote the interests of the chemical industry, engender suspicions about alternative energy sources, and, even while acknowledging their legitimate place in public discourse, trivialize the concerns of ordinary citizens? Although a general curriculum analysis may provoke questions like these, it provides few structures for examining the questions systematically.

PHASE 2: A THEORETICAL FRAMWORK FOR THE ANALYSIS OF *STS* INTERRELATIONSHIPS

Over the last four decades, studies in the history, philosophy, and sociology of science have helped redefine science from objective impartial certification of knowledge to a socially constituted enterprise shaped at many levels by human values, beliefs, and commitments (Collins, 1985; Dickson, 1984; Kuhn, 1970; Longino, 1990; Mulkay, 1975; Ziman, 1984). Sociologies of science, for example, offer insightful perspectives on science, both in practice and as captured in textbooks (see the review by Kelly, Carlsen, & Cunningham, 1993). These perspectives on science provide a conceptual toolkit for the analysis of STS curricula, alerting us to cultural, social, political, and economic forces that shape science, and describe for us how science, in return, has cultural, social, political, and economic features. For instance, both the feminist critique of science (such as that in Chapter 15 by Hilary Rose) and the analysis of the political economy in which science is situated (Dickson, 1984), view science as a product of the culture that produced it.

This conceptual toolkit helped us analyze four of the explicit assumptions found in the *ChemCom Teacher's Guide*:

> 15. The nonuniform distribution of chemical resources on our planet and the demand for a great diversity of goods necessitate a world import-export market economy.
> 28. A social partnership exists between scientific-technological industries and society as a whole. Ideally, such industries deliver useful products and services at a reasonable profit with minimal negative human health and environmental impacts.
> 30. Our present state of knowledge about any given societal-technological issue is likely to contain some imprecision, inaccuracy, and uncertainty. Society must act on the best available information with the understanding that additional information may call for subsequent reevaluation of a given issue or previous solution.
> 16. All sources of information are not equally reliable. The best, most up-to-date scientific information is typically found in professional journals or obtained from scientific associations. (American Chemical Society [ACS], 1988b, pp. xxviii–xxx)

Ripped (somewhat unfairly) from the context of the entire list, these assumptions suggest that: (1) nature dictates a capitalist economy; (2) profits and technology go hand in hand, with inescapable (minor) negative consequences for human health and the environment; (3) we have to make decisions about technologies, but our information is not always perfect; and (4) the best information for decision making comes from scientific experts. Because these claims are assumptions of the curriculum, not claims to be defended or challenged, we view them as glimpses into the curriculum's ideology.

The next phase in our research was to return to the curriculum and examine how the textbook represented the relationships among science, technology, and society.

PHASE 3: SOCIOCULTURAL CURRICULUM ANALYSIS

The third phase of our research, the major focus of this chapter, was to invent and test a new type of curriculum analysis, one appropriate to the study of a curriculum situated in a sociocultural milieu (as are all STS curriculum). In this section we describe and demonstrate what we call a "sociocultural curriculum analysis." The central part of our methodology—a type of content analysis—focused on a single, 64-page section of the curriculum: Unit 5, "Nuclear Chemistry in Our World." Our method is derived from Glaser's (1969) constant comparative analysis, and is designed to provide a complete account of an entire (albeit strategically selected) corpus of text. More traditional forms of content analysis usually analyze only samples of text. For example, Chiappetta, Sethna, and Fillman (1991) compared *ChemCom* to six other best-selling chemistry textbooks. But their analysis was based on a 5% random sample of individual textbook pages. Such methods are inattentive to sociocultural themes, which may be sounded subtly and infrequently, and developed over several pages or chapters. In *ChemCom,* as we will later show, the entire unit on nuclear chemistry is reframed on a single page (American Chemical Society [ACS], 1988a, p. 327) in a philosophically distorted fashion. A 5% sampling strategy would miss that one page, 19 times out of 20!

On the other hand, a curriculum analysis that focuses exclusively on a single unit in a textbook might inadvertently commit a context error as egregious as random sampling. Had we looked at, say, Unit Four ("Understanding Food"), our analysis would have had relatively little to say about the textbook's ideology. The unit poses few economic, epistemological, ethical, or sociological problems; hence, it is relatively unfertile terrain for the investigation of sociocultural influence. We selected Unit 5 because it addresses these problems overtly.

Methodology Outline

• *Step 1: Familiarize ourselves thoroughly with the curriculum being analyzed.* We accomplished this step through a traditional curriculum analysis, as well as through a year-long teaching project in which the first author taught the curriculum in a local public high school, participating in daily planning and debriefing meetings with other teachers.

• *Step 2: Describe and discuss our own values, beliefs, and commitments.* Following Longino (1990), we sought to make explicit our own values. In broad strokes, these values amplify some of the values outlined in the *ChemCom's Teacher's Guide,* but others were added (see Kelly et al., 1993). For example, we view the internal workings of science as much more problematic and socially contingent than the view described in *ChemCom.*

• *Step 3: Attempt a preliminary description of the curriculum's ideology.* By "ideology," we refer to the taken-for-granted (unargued and sometimes unstated) beliefs and assumptions that underlie the curriculum and support the interests of its developer and its sponsors (Anyon, 1979; Mulkay, 1975). Fourez (1989) demonstrated the analytical power of ideological description when he analyzed the scientific literacy espoused by Project 2061 (American Association for the Advancement of Science [AAAS], 1989).

• *Step 4: Identify and archive data.* Given the acknowledged role of our values in our analysis, we archived our data to provide its easy access for others to scrutinize. The primary source text for our analysis is a computer file containing the entire text of *ChemCom* Unit 5, "Nuclear Chemistry in Our World." We entered this text using a computer scanner, and copy edited scanning errors throughout the coding process.

• *Step 5: Code the data.* We used Glaser's (1969) "constant comparative method," which incorporates both theory-based and inductively-generated classifications. Our coding was almost entirely done in collaborative coding sessions involving two or three of the authors, working with both the source textbook and the scanned text file (where the codes were entered). Coding involves labeling source text with brief, analytically significant, categorical descriptions. These codes sometimes emerged from the literature of sociology of science, and sometimes from close attention to the text itself (Strauss, 1987). Coding was iterative; the entire text was coded about three different times (working with both text and previous codes).

• *Step 6 (concurrent with Step 5): Generate rules for coding.* These rules are documented in a coding dictionary and in a series of analytic memos. The rules were successively revised until the analysis reached "theoretical saturation" (Strauss, 1987); that is, additional coding contributed nothing additional to the description of codes.

• *Steps 7 and 8: Classify the codes and write conclusions.* Because our values figure in these steps, we have tried to make these values explicit in our analysis.

Preliminary Description of the Curriculum's Ideology (Step 3)

Our task for Step 3 was facilitated by the explicit enumeration by *ChemCom*'s authors of 32 assumptions underlying their work (American Chemical Society [ACS], 1988b, pp. xxvii-xxx). Although we cannot list all 32 assumptions here, several are quoted below. Four features of the curriculum's ideology became plausible at this stage of our analysis. Each is exemplified succinctly here (see Carlsen et al., 1992, for details).

First, only one of the 32 assumptions, Assumption 20 which claims that engineered technologies improve on natural systems for the benefit of humankind, deals narrowly with the relationship between science and technology. The assumption seems to suggest that science and technology are fundamentally similar enterprises. This conclusion is reinforced by the way science and technology are used throughout the textbook, except for one explicit (and erroneous) definition that says "technology represents the application of science" (p. 78), a hierarchical definition that confers greater status on science.

A second plausible feature of the curriculum's ideology concerns the relationship between science and society. Five assumptions build this bridge. Together they emphasize society's reliance on science. A naive inductivist impression is conveyed through these assumptions. We have already mentioned the assumption that the natural distribution of chemical resources on our planet dictates a capitalist economy. Assumption 13 digs into our legal system: "Societal laws/regulations can either be written in congruence with natural laws or in opposition to them. In the end, nature's limitations and boundaries prevail." What is constructed here is, we believe, a clear hierarchy, with nature, science, and technology at the top, and the unpredictable world of human social behavior somewhere below.

A third plausible feature of the curriculum's ideology concerns the nature of scientific facts and the use of scientific facts in public discourse. Seven assumptions concern the role of scientific and technological knowledge in social decision making. For example, Assumption 8 states, "Expert agreement upon underlying scientific and technological facts related to a given societal or technological issue does not necessarily imply that experts will agree on a particular 'fix.' Social, political, economic, and ethical values influence the opinions and advice of experts." Assumption 9 continues, "Experts sometimes do not even agree on the underlying scientific and technological facts." Although these claims allow experts to disagree about facts, and for experts' opinions and advice to be influenced by social, political, economic, and ethical values, they pointedly do *not* suggest that *facts* are shaped by values. Facts, according to the curriculum, can be *used* in societal decision making, but they have an existence that precedes and maintains independence from social interpretation, evaluation, and negotiation.

A fourth feature of *ChemCom*'s ideology is that responsibility for the (mis)application of science and technology is placed squarely on the shoulders of the voting public. Having established that "All technological benefits are associated with some level of risks/costs/burdens" (Assumption 23), society-specific assumptions relieve science and technology—but not the voting public—of responsibility. "Citizen involvement in public policy is both a right/privilege and a responsibility. Not to decide (or to decide not to decide or act) is also a decision" (Assumption 11).

It is important to recognize the tentative nature of this step in our analysis. These four statements about the curriculum's ideology are not conclusions derived from the textbook. They are a preliminary description of an hypothesized ideology, based on close examination of some supporting materials (assumptions enumerated in the *ChemCom Teacher's Guide*). This preliminary description guided our subsequent systematic analysis in three ways: (1) by suggesting certain classes of codings (for example, codings that concern risk, public decision making, and the nature of facts); (2) by guiding our selection of a unit to focus on; and (3) by making us aware of our initial impressions, and thereby helping us to develop a conscious search strategy for disconfirming evidence.

Coding Data and Writing Rules (Steps 5 and 6)

Our coding and rule-writing of the 64-page nuclear chemistry unit focused on textbook content concerning the sociocultural context of science (see Carlsen et al., 1992, for details). We found that our codes could be grouped into four general categories: (1) rhetorical devices that *de-emphasize the hazards* of technology, (2) rhetorical devices that *legitimate the hazards* of technology, (3) rhetorical devices that *simplify or amplify* an issue (with the three most common strategies being "demonizing," "polarization," and "extremism"), and (4) rhetorical devices concerned with *decision making and control.* Examples of codes in the first and fourth categories are summarized here (see Carlsen et al., 1992, for a full analysis of all four categories).

Category 1: De-emphasizing technological hazards. In Unit 5, we identified five rhetorical strategies that de-emphasize technology-related hazards. The first concerns the use of analogies. We noted two extended uses of sociocultural analogies. In both cases, an STS problem was related to a scenario that is both familiar and comforting. In the first, the text lays the groundwork for discussing radioactivity-related risk by discussing the risks of using different forms of transport to visit a friend (American Chemical Society [ACS], 1988a, p. 311). The data that are provided clearly indicate that the risk of accidental death is 100 times greater for bicycle travel than airline travel. Later, another analogy is used to discuss the disposal of radioactive wastes:

> Imagine this problem. You live in a clean, comfortable home. But you have one big problem. You can't throw away your garbage. The city forbids

> this because it has not decided what to do with it. Your family has compacted and wrapped the garbage the best way possible for more than 30 years. You are running out of room. Some of the bundles are leaking and creating a health hazard. What can be done?
>
> The nuclear power industry, and the nuclear weapons industry in particular, have a similar problem. (American Chemical Society [ACS], 1988a, p. 320)

Because analogies bridge familiar concepts with unfamiliar ones, analogies can be pedagogically strategic and are common in science instruction. Because these particular analogies relate to something comforting (visiting a friend, living in a "clean, comfortable home"), they are ideologically strategic as well. Instead of thinking about the toxic wastes produced by industries and utilities, for example, our attention is shifted to ordinary household garbage in our homes. The analogy does not invite critical questions about the wisdom of generating toxic wastes in the first place; after all, everyone needs a place to live and household waste is inevitable!

Two other rhetorical devices for de-emphasizing the hazards of technology were coded more than once in Unit 5. One is used to emphasize that radiation is "natural" (for example, "The human body naturally contains a small amount of radioactive material", p. 267); another suggests that other countries have less restrictive attitudes and regulations with respect to radiation. Given America's preoccupation with international economic competitiveness, such references may do more than simply suggest precedents; they hint that if we drag our technological feet, we may be left behind.

We need to emphasize that *ChemCom* is not relegating technological hazards to the null curriculum. The text explicitly lays out many of the risks and dangers associated with radiation. This is, in our judgment, a tremendous improvement over virtually every other mainstream chemistry (and physics!) textbook on the market. But careful attention must be paid to how issues are framed. Although technological hazards are acknowledged in *ChemCom,* we found evidence that they are also deemphasized through subtle rhetorical strategies.

Category 4: Decision making and control. The fourth general category of rhetorical decision-making strategies shapes the reader's perspective on how STS decisions are made and should be made. Three examples follow.

Presenting a *scientistic view* is a rhetorical strategy that (erroneously) suggests that science can provide unproblematic solutions to complex problems. Overt instances of this strategy in *ChemCom* are uncommon; in fact, some of the curriculum's assumptions explicitly refute the idea that science provides simple answers. But consider the following passage, which opens the chapter, "Living with Benefits and Risks":

> In the following sections, we will examine how to balance for ourselves the benefits and risks of any new technology. One must weigh the benefits of the technology against the number of people who might be seriously harmed or killed by it. *Statistics and logic would define as acceptable those new technologies in which the probability of risk is low or the benefits outweigh the risks* [emphasis added]. (American Chemical Society [ACS], 1988a, p. 310)

Statistics and logic do nothing of the kind. People may choose to make such decisions, but even the most quantitative analytic methods reflect the values of their users. *ChemCom* discusses the risks (and even the costs) of new technologies in many places. Unfortunately, formulations like the passage above may distort readers' perspectives on "rational" deliberation.

The culminating activity in Unit 5 is entitled "Putting It All Together: Separating Fact from Fiction." The activity divides students into groups and asks them to revisit the "Survey on Nuclear Phenomena" that began the unit. We coded the methodology for accomplishing this as *philosophically naive,* because it commits at least two fatal philosophical errors that are likely to figure in students' decision making. The activity asks students to classify the 40 survey items into the following categories:

> *Scientific reality* (S). The item is either scientifically correct or incorrect as stated. It reflects a principle that can be verified or disproved via observations in a clear-cut experiment.
> *Political reality* (P). Data published in reputable journals, magazines, or newspapers would support the notion that the item reflects (either positively or negatively) the public opinion or current situation.
> *Value-laden statement* (V). The item contains emotionally charged words and reflects a definite pro- or antinuclear stance. Note it is possible for such an item to be either based on, or in contradiction to, some scientific principle. (American Chemical Society [ACS], 1988a, p. 327)

The first category is philosophically untenable. The myths that science definitively verifies or refutes facts through clear-cut experiments were debunked decades ago by Karl Popper (1959) and Imre Lakatos (1970), respectively. The third category is also philosophically untenable. It suggests that statements can be purged of values by either deleting emotionally charged words or by communicating an image of balance. Taken together, the two erroneous definitions reinforce *ChemCom*'s hierarchical view of the epistemic status of scientific and nonscientific knowledge: scientific facts are "hard," values are inevitably "soft."

CONCLUSIONS

ChemCom itself eschews the use of emotionally charged words, and it effectively communicates a perspective of balance. The book engages in no environmentalist bashing, for example, and does an excellent job of pointing out the real and perceived risks of chemical technologies. Nevertheless, the curriculum is thoroughly value-laden, as we have demonstrated in our analysis. Perhaps this is an inevitable consequence of enhancing student interest by dealing with the social context of chemistry (John Ziman's "relevance" approach, Chapter 3). More troubling is the fact that the curriculum's values are sympathetic with, if not identical to, the values of the American Chemical Society's numerous corporate affiliates, who are not mentioned in the textbook.

In Step 3 of our analysis we provisionally identified four features of the curriculum's ideology. First, we suggested that the curriculum's authors see science and technology as fundamentally similar (if not hierarchical) enterprises. Second, we suggested that the curriculum holds a hierarchical view of the relationship between science and technology and society, with the former holding clear implications for the latter, in economics, law, and other human endeavors. Third, we proposed that the text vested in scientific facts special (and unwarranted) ontological status. Facts, according to the curriculum, can be used in decision making, but have an existence that precedes and maintains independence from social interpretation, evaluation, and negotiation. Fourth, we suggested that the curriculum placed responsibility for the (mis)application of chemical technologies squarely (and possibly unfairly) on the shoulders of John and Jane Public, rather than, say, at the feet of chemical industries and their shareholders.

Our analysis of Unit 5 lends support to our preliminary interpretation of the curriculum's ideology. The text de-emphasizes and legitimates the hazards of technology, simplifies and amplifies complex issues, and distorts the nature of decision making and control related to science and technology. These goals are accomplished through rhetorical strategies that effectively bury the authors' point of view beneath a veneer of balance.

When teaching *ChemCom* in a public school (in 1991–1992) and discussing the curriculum with preservice and experienced science teachers, we observed that the textbook's ideology is not at all evident to its readers. In general, students and teachers see the textbook as providing a balanced, objective view of the role of chemistry in societal decision making.

Although we find an uncritical acceptance of the textbook's explicit and implicit messages distressing, we should probably not find it surprising. Most science textbooks assiduously avoid discussion of controversial issues. Consequently teachers and students are not used to evaluating their textbooks critically. Discussions about the social consequences of science and technology in textbooks are usually brief, and often, we suspect, overlooked. On the other hand, when "educational materials" from special interest groups make their way into the science classroom, their messages are often obviously one-sided and easy to dismiss.

In this chapter, we have tried to make clear our critical stance through our selection of examples, our explication of our values, and our intentional use of words with mildly pejorative connotations, such as "ideology." We cheerfully acknowledge that in doing so, we have left ourselves vulnerable to some of the same charges we have leveled at *ChemCom,* such as the accusation that rhetoric is used to simplify issues that are actually complex. This is, we believe, a risk worth taking, in order to stimulate further conversations among teachers, educational scholars, science content specialists, and curriculum developers.

We do not believe that *ChemCom*'s rhetoric is the product of a secret conspiracy of chemical industry propagandists. A conspiracy is not needed to account for the problems in the text. Although we disagree with the curriculum's underlying ideology, we recognize its pervasiveness and acknowledge that reasonable people believe in it. *ChemCom* presents a view of science that is socioculturally naive, but so do most textbooks (Anyon, 1979) and undergraduate science programs (Bruffee, 1992; Tobias, 1990).

When all is said and done, we are actually quite fond of *ChemCom.* We believe that approached critically, *ChemCom* provides a rich resource for science teachers and students alike. Yes, we can use the textbook without being used by the textbook.

PART III

PUBLIC UNDERSTANDING AND CULTURAL DIVERSITY

One of the challenges to the STS movement voiced by David Layton in Chapter 4 was to consider the impact of STS science on national cultures. "To what extent," he asked, "is [school STS] premised on a model of participatory democracy and on assumptions . . . which do not readily transfer to all other contexts?" In Part III this challenge is addressed.

We begin in the industrialized world where scientific literacy, or public understanding of science, has a high profile. Chapter 10 claims that being able to discuss scientific issues about risk and about the environment is central to this matter. Immediately knowledge becomes coupled with values. The public needs education, both formal and informal, but moral, religious, and ethical considerations are also essential to reach any personal conclusion. Television and museums are considered by Joan Solomon (Chapter 10) and Arthur Lucas (Chapter 11) not just for the scientific information they give, but for the way the public interacts with them.

Many African countries express dissatisfaction with the imposition of imported Western technologies. Olugbemiro Jegede (Chapter 12), like others, looks to STS education to provide a basis for appropriate indigenous technology. At the same time he recognizes the cultural influences on children's learning about the environment. Traditional worldviews, societal structures, and local beliefs about plants and animals, all play a part. Despite the conflicts that these sometimes produce, the author urges that traditional cosmologies should be incorporated into STS education.

Parts of India exhibit similar features. Here again there is an ancient culture that sometimes comes into conflict with a scientific perspective. In the last chapter of this part of the book, Anita Rampal describes the action research of a class of young girls involved in a local STS program. They use an experimental approach "to alleviate the sense of alienation normally associated with science." It requires "patience, understanding, and empathy to present an alternative world view without causing offense." Her girls do this, while acting as participating citizens in the face of a possible health risk to their community.

Chapter 10

Knowledge, Values, and the Public Choice of Science Knowledge

Joan Solomon

In many countries there is a growing movement for measuring the public's understanding of science (for example, Durant, Evans, & Thomas, 1989; Miller, 1989). Every 2 years, the United States produces science indicators for this purpose. In Britain the Royal Society published a paper on this theme in 1985 that triggered a series of related research projects. Similar work is underway in Europe.

This whole subject is of central interest to the STS movement. STS educators have always repudiated the idea that scientific knowledge is, or should be, essentially esoteric or exclusive, and that only experts can make decisions relating to the impact of science on society. However, the two counterclaims of STS (the public accessibility of regular hard-core scientific knowledge and the public participation in decision making) may describe two rather different kinds of scientific learning. The first requires the scientific knowledge that scientists, like science teachers, claim should be a part of everyone's education so people can "get it right." The second concerns more personal and problematic knowledge whose purpose is to evaluate public dilemmas where affective and moral reactions will also be concerned.

In his early analysis of scientific literacy, Jon Miller (1983) claimed that there were two dimensions of scientific knowledge. He referred to them as: (1) being learned about science, and (2) being able to give an opinion about some scientific matter (p. 29). Later Miller (1989) subdivided the first category into two parts: an understanding of the processes of science, and a vocabulary of scientific terms and concepts. The essential difference between the two main categories remains, however.

Acquiring scientific literacy in the first sense is a relatively straightforward matter within the school, and is not difficult beyond school if such knowledge is actively sought. There are adult information courses in science that may provide access to further education of a formal kind. There are also a variety of informal sources of knowledge available to the public, as documented by Lucas (1983). Many of them, such as museums, libraries, science television programs, and interactive science centers, are almost as valuable as formal science courses for fulfilling Miller's first aspect of literacy.

The question facing researchers into the public understanding of science and STS teachers in the schools is how to address Miller's second aspect of scientific literacy. Can there be teaching that enables learners to give their own opinion? The lack of any knowledge sources specially designed to fulfill this need has rarely attracted comment. It has simply been assumed that the two kinds of science sources were coterminus. Either being learned about science would enable the person to give an opinion, or needing to give an opinion would just entail learning some more straight science.

However, the question is further complicated by the differences between adult and student learning. When science knowledge is brought to those outside the school, normative approaches no longer work, and science knowledge will be both sought and received in a different spirit, for at least two reasons. In the first place, members of the public are often closer to the interface between science and society than are students in the classroom. The public may have a personal need to decide about a medical treatment or to consider how to cast their vote on some science-related issue. Secondly, we need to remember that adults expect to be able to exercise complete choice over what they will or will not learn. Their motivation will not concern passing grades, and may not even include cognitive satisfaction; it could be capricious or deeply emotional. Therefore, Miller's first category of being scientifically literate will be achieved via a number of different routes in different personal situations and will be driven by a number of different motivations. The knowledge that the public acquires may be colored by these different pathways of acquisition in a way that may make that knowledge much closer to social issues and people's opinions about them.

STS teachers would also like to see their students' scientific knowledge being used in the context of social issues. However, knowledge gained by the very formal methods in schools may not be easily transferred to a discussion of social dilemmas. In an early study of how young students used their school knowledge of energy and its conservation in a discussion of public energy dilemmas, it was shown that the formality of school science knowledge did, at first, inhibit student discussion (Solomon, 1985). The path between understanding conceptual science and expressing an opinion about it (Miller's first and second categories of literacy) is not linear in school or in the public domain.

This chapter will examine how far these two dimensions of scientific literacy coexist. It will refer to work carried out:

1. In the public domain where there is freedom of choice whether or not to learn science, but also where there are risks and concerns which may introduce strong affective and moral factors
2. Within school science where there is prescriptive learning that is hard to avoid, but where there is some shelter from adult problems and responsibilities.

STYLE CHOICE AND INFORMATION SOURCE

There are more and less subtle ways of offering science to the public. At one end of the scale is life-world knowledge within language and metaphor, where the science component has become familiar through long usage. By this process phrases like "harnessing power" or "feed a cold and starve a fever" have become so deeply embedded in the social stock of knowledge that they have long since shed their science label. How this may have come about has been worked out in some detail for the case of energy and its technologies (Solomon, 1992a). As this knowledge lost its science affiliation, it also became inescapable within general talk. Maxims and familiar language are not chosen for original knowledge content, only for immediate convenience. Science knowledge that is still fresh from the knowledge-making process is often considered too obtrusively academic for general conversation.

There are plenty of public information sources for students and others alike. Studies of school students in the United States (Wiesenmayer, Murrin, & Tomera, 1984) and in Britain (Solomon, 1992c) have shown that some students claim to get most of their knowledge about science-based social issues from television, some from school, some from reading, and yet others from talking with friends. They might also have included radio, museums, botanical gardens, and hands-on science centers. There is great variety of media. How do people decide whether or not to receive this knowledge?

Some research has investigated people's preferred style of information reception. One ingenious approach was to consider image construction in the different modes of communication (Wober, 1985). This research seemed to show that those who learn best from books or radio do so out of a preference for constructing their own images. Quantitative studies suggest that this imaging preference is far stronger than the contrast between the visual and aural media (books and radio) because there is a substantial overlap between the populations that use them. There is much less association between those who prefer learning from books (or radio) and others who prefer television. Although Wober (1985) and Marks Greenfield (1984) imply that this is merely the reverse of the imagery choice, there are other data to suggest that learning from television correlates quite highly with claiming to learn best through talking with other people. This is no chance association. Communication studies have long indicated that television reception has a great deal to do with personal responses to screen characters and television personalities, a phenomenon sometimes called "para-social interaction" (McQuail, 1984). They have also shown that adults and children alike often need to discuss what they have heard or seen as an essential part of the reception of knowledge (Hodge & Tripp, 1986).

Recent research carried out in the United Kingdom further explored the link between viewing television programs about science-related issues and discussion of them with friends. In the *Discussion of Issues in School Science* (DISS) project, small groups of 17-year-old students tape recorded their informal discussions of television excerpts about controversial issues (Solomon, 1992c). They also answered a short questionnaire about how they had most often heard about "science issues"—a term defined by examples in the questionnaire.

The two sets of data showed less association than had been expected. Those who most often referred to television programs were not the same ones who answered on the questionnaire that they learnt most about science issues from television. There was a similar anomalous result for reading. This lack of correlation between attitude responses and behavioral outcomes is not a new phenomenon. Eiser (1986) contends that the lack of correlation between attitudes and behavior is a common finding. He added that little notice seems to have been taken of it. "Nonetheless such results have generally failed to shake the conviction on the part of most attitude-theorists that attitudes are an important, if not the major, cause of the kinds of behavior which interest social psychologists" (p. 53). The results from the DISS project may be showing that people have a sense of their favored style of knowledge reception, whether or not it is borne out in practice by their behavior. From the point of view of public understanding, these results provide evidence for the curious and noncognitive nature of people's uptake of science knowledge. From the research perspective, these results cast considerable doubt on the use of the attitude questionnaire as an instrument of research into views about STS. This point has been made trenchantly by Aikenhead, Fleming, and Ryan (1987).

In addition to people's preferred style of learning there is also a simpler choice people make of whether or not to learn about science at all. People have easy control over the television set. There may be museums, hands-on centers, or other science shows, but members of the public do not need to attend them. If it is not to their liking, they stay away (except if they have young and persistent children). Even then people have a skilled repertoire of avoidance mechanisms (such as refusing to read explanatory labels) that they can use for protection against formal science knowledge. Sources of conceptual science knowledge may be available, but not everyone sees the point in using them.

KNOWLEDGE AND RISK

To investigate the ability of the public to express considered opinions, it will be necessary to find a topic that touches people closely, such as promising benefit or threatening personal risk. One such study of public understanding was related to a British radon survey in private houses. This was carried out by Michael (1992) who found that people were collaborating with his survey, showing interest in the findings that were potentially hazardous to them, but not wanting to learn about the science concepts involved. Michael wrote: "Certainly there is a case to be made for saying that the speaker situates him- or herself at a considerable distance from science" (p. 318). Michael made a sharp distinction between the public's interest in two different kinds of science knowledge: science-in-general (conceptual ideas) and science-in-practice (applications and implications). Members of the public could agree with the intentions of the radon survey, and yet continue to profess a complete lack of interest or understanding of its science concepts. These two categories of science knowledge show strong similarities with those of Miller (1983) quoted above.

Until recently, there was little evidence to confront the eminently sensible belief that when people really need to understand science (for example, when they are at risk), they will make an effort to do so. Now, it seems, we know better. From studies by Jenkins (1990) of pensioners and their thermometers, by Macgill (1987) of residents of Sellafield where the British nuclear reprocessing plant is situated, and by Lambert and Rose (1990) of those with congenital hypercholestrolaemia, we can see how little the public appears to be concerned to learn more about the scientific knowledge relating to their predicament. In the first two studies mentioned, as with a study of Sellafield apprentices by Wynne (1990), it is easy to understand the incentive to refuse knowledge. Thinking about the science that underlies some unavoidable personal risk just increases anxiety. This produces what Wynne has called the "ignorance is bliss" syndrome.

By contrast, the medical case study reported by Lambert and Rose (1990) concerned a risk that appropriate knowledge does make avoidable.

The congenital disposition to high levels of cholesterol is life threatening. Both those who are born with it, and their parents, have great need of information. The research showed that patients actively interpreted information given by their doctors, learned the skills necessary for identifying foods that contain cholesterol, and gained a quite sophisticated working knowledge of different fats. Yet very few patients were either "aware of, or displayed any interest in, the scientific details of their metabolic condition which might make sense of their condition" (p. 10). Even in the face of possible premature death, they still exerted their choice *not to understand* science concepts. This links nicely with Michael's (1992) conclusion that lay people may show interest in and commitment to the applications of scientific knowledge, and yet distance themselves from the principles that explain those applications.

DISCUSSING SCIENCE

To examine the kind of scientific literacy that enables people to express an opinion (Miller's second category), we need to set up a situation in which (1) people feel able to discuss together, (2) a worthwhile controversial issue is presented, and (3) there is access to knowledge of various kinds relating to the issue.

STS education has a commitment to the free discussion of issues relating to science. STS is more than a movement for relevance in science, it takes on board all the concommitants of relevance. As Ziman (1980) wrote about STS for 16- to 19-year-old generalists: "The real need is for an integrated course about science in all its various aspects—historical, philosophical, cultural, sociological, political . . . dangerous, equivocal, ethical etc." (p. 143). Although traditionalists have expressed reservations about whether science teachers can or should attempt to include ethical points of view in their lessons, most STS teachers and curriculum developers have few doubts about it. This is not because of some special arrogance about their competence. After all, every parent teaches everyday ethics to their children without a certificate in the subject. Nor is it just because STS educators want to present science as a human activity that affects us all. The objective is more subtle: it derives from a concern about the public perception of science. This perception has suffered a great deal from a remote elitist image conveyed by science, and from science's association with weapons, warfare, and the use of animals in experiments. In the face of an image of scientists as uncaring and amoral, it will not do to banish the discussion of ethical issues from the science classroom to some other lesson in personal or social studies. That would only serve to confirm the negative public image of science. The STS teacher and an STS curriculum strive to encourage the kind of debate where the students and teacher can use their feeling for what is right, as well as their knowledge of science, in the normal context of learning science.

Research carried out in the DISS project (Solomon, 1992c) casts some light on how these two extremes of scientific literacy—scientific knowledge and informed discussion—may fit together. The number of occasions in which the students' exchanges overtly related to scientific knowledge was small. The transcripts of discussion were coded for any passages where one student instructed the others in some area of scientific knowledge. In a few cases where a group of students said that they did not understand the language of science, the discussion never got started. The students would say things like, "Why don't they speak English?" and "They try to blind you with science," and then the talk stopped. In all other cases the technical terms (such as "catalytic converters" and "isotopes") passed without comment and discussion proceeded. This suggested that some simple familiarity with the terms of science, rather than Miller's first characteristic of literacy (being learned about science), was a tacit prerequisite for Miller's second characteristic of literacy (being able to give a reasoned opinion).

These findings show that science knowledge is not irrelevant to affective discussion. This is an important result. STS has always included controversial issues about which students might have deeply felt reactions. More traditional teachers may be worried about this and fear that the affective will drive out the cognitive. Now that there is some firm evidence that this is not the case, the link becomes stronger between STS teaching in schools and the concerns of the public. The research findings also mean that STS teachers need to take into account not just their students' preconceptions about science topics, but also their preconceptions about ethical positions.

ENVIRONMENTAL ETHICS AND *STS*

In the last 10 to 15 years, environmental issues have become central to the public's understanding of science and to their concern for health, for wildlife, and even for the future of the planet. Young people often feel most deeply about these matters. Social and moral dilemmas associated with environmental issues are proving controversial and have achieved prominence in the media programs. Listening to these programs gives grounds for believing that the environment is presenting the public, young and old, with new challenges to old moralities.

British schools are urged to provide environmental education that will help students to clarify their values on controversial social issues. In practice, however, there has been a sad failure to marry the two strands of curriculum thinking (the moral and the environmental) in the new National Curriculum for science in the United Kingdom. Even within personal and social education there is no guidance from research. Environmental issues did not figure at all in Kohlberg's (1984) dilemmas, which he used to probe the moral development of the young, and were only peripherally involved in the discussions of the DISS project. Indeed, the only substantial British work on young people's moral development with respect to environmental reasoning of which I am aware is to be found in recent work by Haste (1993). Here she argues from her data that "moral vision" could either be the product of a moral stage of reasoning (Kohlberg & Candee, 1984) or of a much simpler schema of right and wrong with respect to observed suffering.

The important educational task of helping our young people to use their scientific knowledge (when considering moral questions in relation to the natural environment) is made all the more difficult because of a general lack of structure in the whole field of environmental ethics. This lack is due to the diversity and changes in public attitude, and also to a failure on the part of most philosophers to reconsider various positions.

The classic philosophical work by Passmore (1974), *Man's Responsibility for Nature,* examined the Judeo-Christian traditions of "domination" and "stewardship" toward the environment with respect to modern issues such as pollution of air and water, its effect on wildlife, and people's duty to care. He concluded then that no new ethical approach was required. This was in contradiction to the earlier American seminal arguments by Leopold (1949) on the need for an attitude of moral obligation towards the whole biotic community, including the biosphere. Passmore's position was reconsidered, in the most part favorably, by Attfield (1983) and Taylor (1986). However, it could be claimed that the whole philosophical climate of thought has changed radically in the last few years.

There are substantial problems in continuing to apply the notion of stewardship toward the environment. Some authors, such as Roszak (1972) in the United States and Paden (1992) in Britain, have urged a mystical or transcendental attitude towards "Nature" which could replace God as the steward's employer. This might have been a prevailing attitude of the hippy "flower-power" years, but there is doubt if it has any great public or academic following now.

Another approach to the stewardship problem is to continue in a "management of rural resources" tradition. A great deal of what we now know about ecology has arisen out of the craft knowledge of wardens and gamekeepers. These were stewards in an almost literal sense, and more and more of our young people earn their living in jobs where special environments—woods, national parks, rivers, and even landscaped tip sites—are expertly managed. According to Wooster (1977), the early advances of ecology grew out of this management business. It is also interesting to note that the tone and language of the scientific terms coined for ecological concepts in the 1930s (for example: "producers," "consumers," and "energy flow") are strongly reminiscent of the Ford production line of this era. Stewardship was being taken very professionally!

Today we have the threat to us all from ecological damage, such as the thinning of the ozone layer. These threats have made an impersonal approach to the environment almost impossible to sustain. We are not so much stewards now, as potential victims, along with all other species. The special place of humankind within creation has been further eroded by the progress in ecological thinking that claims there can be no privileged positions within any ecosystem. All organisms react upon each other. All life is interdependent. Finally, the "Gaia hypothesis" of the biologist James Lovelock (1979) has suggested that the planet itself could recover its equilibrium state after almost any ecological disaster, even if it put an end to humans and all other life forms. This may describe the earth as a self-equilibrating system, but it certainly does not put it into a semidivine position.

Under these circumstances it may be salutary to recall the universal reaction of the astronauts who entered space in the 1960s and 1970s. Their reaction was not awe in the face of the black emptiness of the void, but a sense of marvel and delight at the unique beauty of earth. When we combine these instinctive reactions of the astronauts with an understanding of the scientific principles of ecology, it seems as if some new relationship between humans and their environment may be in the making. Although it is certainly rash to make any hard predictions about such a new program, all STS teachers will have noted the strong sense of protection and kinship that children feel for animals. This probably springs from children's own feelings of powerlessness in the face of the adult world.

The rationale of STS teaching has emphasized the different concerns of different societies. It will be important to recognize, therefore, that environmental values may have a strong cultural component.

CULTURE AND ENVIRONMENTAL VALUES

Native peoples have often claimed a special relationship with the environment. This has acquired some prominence in recent debates on environmental issues. Booth and Jacobs (1990) write of the Native American belief in the need for "reciprocity and balance" with nature. They give a striking quotation from Chief Standing Bear writing in 1933 about the Lakota people of the great plains, which fits rather well with trends in environmental ethics noted in the last section. "We are of the soil and the soil is of us. We love the birds and beasts that grew with us on this soil. They drank the same water and breathed the same air. We are all one in nature" (p. 28). Whether or not such a vision is colored by nostalgia for a bygone age, it is clear that STS teachers would have their task made easier if they were working from within such a strong and uniform cultural tradition. Such people could have no doubt about the connections between environmental science and their society. It is only when native children are taught a science that is a cultural import from a society no longer living close to nature that the STS teacher notices intellectual and moral conflict.

There are a variety of ethical positions that students and their science teachers may hold. This can and probably does lead to conflicting situations within the classroom, but the STS teacher does not dodge controversy. Instead, such teachers use it sensitively, and with the addition of relevant science knowledge, to encourage their students to discuss issues and determine their own position. In order to do this well it is clearly important for teachers to appreciate the range of environmental beliefs their students may hold.

How diverse these beliefs can be was shown by a small pilot study I carried out among student science teachers ($N = 45$) and 13-year-old students ($N = 57$) in several British schools. The respondents arranged four statements in ascending order of importance according to how closely the statements match their own views. The results in Figure 10.1 show a mixture of attitudes toward environmental responsibility. For most of the student teachers, the whole-planet ethic (statements 6, 7, and 8) was stronger than human-centered obligations (statements 1 and 2), and also stronger than a stewardship attitude toward wildlife (statement 3). The students, by contrast, showed the expected concern about wildlife and future generations (statements 3 and 6).

Although too much weight cannot be placed on such a simple questionnaire methodology, the results indicate a sense of moral obligation that is in urgent need of study. Research should help teachers design and deliver environmental education in which their own feelings of responsibility are unlikely to match exactly those of their young students.

SUMMARY

This chapter has linked current interest in the public understanding of science with school STS education. The two have a great deal in common

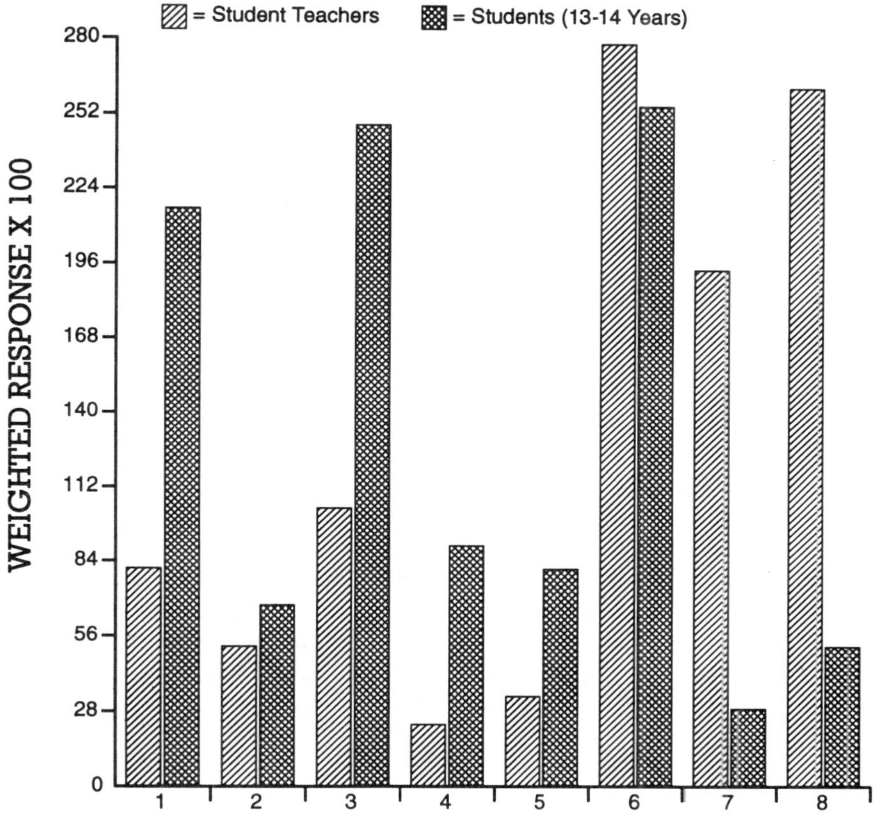

1. We must take care of the environment for the sake of our health and wealth.
2. It is our moral duty to care for all of God's creation.
3. All wildlife in the environment has rights we must respect.
4. We also have moral obligations towards our trees.
5. We need to preserve unspoilt natural areas for our own recreation.
6. We must take care of the environment for the sake of future generations yet unborn.
7. The ecological balance needs to be carefully managed to preserve our natural resources.
8. Our planet is a living entity which we must respect.

FIGURE 10.1 *Responses to the question, "What is your attitude toward the environment"*

with respect to content. Research shows that members of the public who receive information from informal sources accept it most readily when they feel able to act upon it. On the other hand, the public often shows little interest in acquiring conceptual knowledge. STS education aims to deliver scientific knowledge of a kind that is relevant and useful to students by helping them to apply it in the social context of the everyday world.

The themes we find in most STS courses are crucially important for society. Some involve public controversy and ethical dilemmas. With the high public interest in environmental protection, a substantial proportion of this chapter has been devoted to the consideration of ethical positions relating to the natural environment. This is a complex area where a number of views are held, sometimes with deep commitment, and where different cultures and even different age groups may see their responsibilities in very different ways. This variety brings us back to the importance of choice and values in STS instruction.

In traditional science instruction, personal opinion is not involved and may be actively avoided. STS instruction, on the other hand, seeks out exchanges between students to help them arrive at personal positions that combine scientific knowledge with moral responsibility. Evidence is beginning to show that students can do this. In terms of Jon Miller's (1983) definition of scientific literacy, students are acquiring learning about science and, at the same time, the power to express an opinion on a scientific matter that has social and moral dimensions.

Chapter 11

STS Beyond School: Public Perceptions and Sources of Knowledge

Arthur Lucas

When people leave school, there are many opportunities for them to encounter scientific information. I have discussed these sources and their educative role elsewhere (see particularly, Lucas, 1983, 1991).

The press has some coverage, even in the tabloids, of major scientific issues that impinge on health, crime detection, or, increasingly at present, on environmental issues. The nature of that treatment in the news pages can vary from newspaper to newspaper, with the major variation in the amount of background information provided to the news stories. Some newspapers have a major commitment to science reporting. *La Vanguardia* in Barcelona, for example, has a weekend "Ciencia y Technologia" supplement, with some articles that are *not* about applied science or science with utilitarian ends. The issue of November 18, 1989 had a major article on insectivorous plants. That of March 3, 1990 marked the 150th anniversary of the death of the German amateur astronomer Olbers by considering the origins of his question, "Why is the night sky black?" and discussing the cosmological implications of the search for an answer.

Many feature articles in the quality press discuss the implications of scientific and technological development on the life of individuals and society. For example, a feature article in *The Independent* (Lloyd, 1990) reports the identification of the gene that produces an enzyme, iduronate sulphate sulphatase, whose absence produces one of the mucopolysaccharide diseases, Hunter's syndrome. As well as providing a discussion of the causes of the disease, the article offers an explanation of X-linked genetics, and a very brief account of the methods of "proof" used to identify the gene. There is a matter-of-fact statement, "it will be four or five years before it is possible to help sufferers by injecting them with the gene that they either lack or possess in a damaged form" (p. 19). There is no hint in the feature article of the controversies surrounding gene therapy. In the news article on the same research in the same issue of the newspaper there was the comment, "in the future, it may be possible to treat affected embryos with the gene that they lack, or which is defective, though gene therapy raises difficult ethical issues" (Lloyd & Hall, 1990, p. 3). But in the comments on the other aspects of the identification, such as "there is the real possibility that we can screen early in pregnancy" (p. 3), there is no hint that other ethical and moral questions are also present; for instance, does agreeing to a screening for the presence of a genetic defect that will produce a major disease imply at least tacit acceptance of the abortion option?

Television documentary and news programs share similar problems. What is to be presented? What "story" is to be told? Will the absence of controversy make "good" television? (See Silverstone, 1985.)

Museums are also thought to be a potent source of knowledge of science. The direct connection between the aims of at least some of those in the science center movement and the aims of STS can be seen in the words of Frank Oppenheimer, the founder of the Exploratorium in San Francisco that provided the model for most of the existing centers around the world:

> If people feel they understand the world around them, or probably, even if they have the conviction that they could understand it if they wanted to, then and only then are they also able to feel that they can make a difference through their decisions and activities. (Oppenheimer, 1985, p. 8)

An intimate relation between values and conceptual knowledge is clearly assumed here.

However, most science centers in the United Kingdom that set out to present science concepts make scant reference to the implications of the application of those concepts. The underlying science and the social implications are rarely both made explicit in the same exhibit. For example, the "grain pit," one of the most popular exhibits in the Science Museum's "Launch Pad," shows methods of transporting solids: an auger, a conveyor belt, and a bucket dredge. Nowhere in the exhibit itself is attention drawn to the application of the principles of the inclined plane in the Archimedian screw that comprises the auger, to the principles of mechanical advantage, or to gear ratios. Neither is there any discussion of the influence of mechanization on the social life of employees. What visitors learn is problematic: do they leave with an idea of the use of specific mechanisms? Do they make any generalizations? If so, are they appropriate ones, even as simple as "solids can be moved by a variety of mechanical methods?" Or do visitors make erroneous assumptions, such as "in flour mills wheat is moved by men turning handles to drive augers and conveyor belts?" In other words, are visitors able to recognize *which* facets of the exhibit model the use of these devices in industry?

This chapter explores STS education from the perspective of education outside the school classroom. I begin by summarizing the results of research into the public understanding of science and into the relationship between this understanding and the public's position on science-related policy questions. The research poses challenges to STS science educators who aim to improve students' interpretations of, and decisions on, science-related controversies. The chapter concludes with a fundamental implication for STS education in the school curriculum.

PUBLIC UNDERSTANDING OF SCIENCE

It is important to recognize the ambiguity in the term "understanding of science." The term is used to describe both knowledge of scientific concepts and knowledge of science as a sociopolitical phenomenon. It is the difference between learning *science* and learning *about* science.

Early studies of the public understanding of science were concerned with the views of the public about science. A series of studies conducted by Miller in the United States has tracked the attitudes of the U.S. public towards science (for example, National Science Board, 1988). Similar studies have been conducted in Europe (Commission of the European Communities, 1977).

Knowledge of Scientific Concepts

Initially, the level of public knowledge of scientific concepts was studied very indirectly, using questions such as, "Do you understand the meaning of the term radiation?" (Miller, 1983). However, more recent studies have included questions more directly assessing the knowledge of the respondent. British examples include Durant and colleagues (1989) and the series of papers reporting aspects of 1986 data collected for me by the public opinion polling organization MORI (Lucas, 1987a, 1987b, 1987c, 1988, 1989). (My 1986 survey of the public knowledge of science was supported by a grant from the Nuffield Foundation.) Although not assessing the same sample of scientific concepts, aggregated scores on the knowledge scales derived in each study present similar pictures. In both, the typical respondent was able to answer about half of the simple questions.

Knowledge of the Nature of Science

It is much more difficult to assess people's knowledge of the nature of science than it is to see whether they have at least some grasp of a scientific concept. The topic area is complex, and it is at least arguable that there is not a homogeneous "science" that can be assessed by asking methodological questions such as:

> What is the best description of the way that scientists work?
> 1. Careful collection of data which are then sorted and compared with other data to establish relationships between variables.
> 2. Designing experiments to test theories or hypotheses, by comparing the actual results with predictions.

Not only is the nature of this question problematic (are we asking a sociological question, describing how scientists have been observed to work in studies of "laboratory life" or biographies, or are we seeking a normative answer justified by philosophical analysis?), but the question's either-or form assumes a homogeneity of scientific practice or norms (can the methodology of biology be the same as that of physics? See Rosenberg, 1985, for a discussion). Answer 2 does not admit the geomagnetic studies that enabled the influence of the sunspot cycle on magnetic fields to be detected in the last century, and excludes fields as diverse as electronmicroscopy and geological mapping. Answer 2 also conceals the issue of the idealization of scientific concepts, implying an empiricism that sees real-world results as the touchstone of truthful evidence, ignoring the inconvenient experience of, say, actual pendulums, none of which truly obey the "law of the pendulum." (See Matthews, 1990, pages 34–38 for a discussion of the idealization issue in the context of classroom science.) Answer 1, on the other hand, can imply acceptance of completely theory-free naive inductivism, rather than the more sophisticated collection of data in a "Humboldtian" program (Cannon, 1978) that informed early (and to an extent, current) research in meteorology, geophysics, geology, epidemiology, and biogeography.

Notwithstanding the difficulties of assessing philosophical and sociological aspects of knowledge of science, attempts have been made. Durant and colleagues (1989) used a screening question to establish whether respondents felt that they had a clear or a general sense of what it means to study something scientifically, and then asked them, "What does it mean to study something scientifically?" Less than 14% of respondents mentioned theory construction and hypothesis testing or aspects of experimental method. However, asked to choose between possible approaches to checking a suspicion that a drug used to treat high blood pressure was not working well, 56% chose the response involving a controlled experiment. This result suggests that the level of tacit understanding of the experimental methodology was higher than indicated by the open-ended question. Similarly, while only 4% mentioned theory building in the general question, 55% disagreed with the assertion "*All* of today's scientific theories will still be accepted in a hundred years' time."

In my survey (Lucas, 1989), we asked respondents to say whether they thought that each of several phrases, such as "Find patterns in results from experiments," "Eliminate hypotheses," "Look for results which do not fit their theories," described what scientists did. Although I am not entirely satisfied with the quality of the data from this question (because we were not able to probe the meaning attributed to terms such as "observe," "patterns," "experiments," and "control"), the responses indicate a general consistency with a naive inductivist view of science (a strong belief that scientists measure, make observations, and use observations to make new theories). There is less support for the refutationist view implicit in seeking results that do not fit theories. Among the more scientifically knowledgeable, however, the support for eliminating hypotheses is suggestive. Perhaps the more knowledgeable group has a mixed model of scientific methodology? Although derived from a different type of question, the data are not inconsistent with the Durant study.

No one has explored the views of the public about the historicity of science and the relationship between theories that become superseded and those that replace them. Neither have we explored the views of members of the public about the way that theories are replaced. How many would support the Kuhnian position on theory transformation summarized by Matthews (1990, p. 45) as depending "not so much upon rational persuasion as they do upon mob psychology and the mortality of the aged"? How many would share the view of scientific discovery as a discrete event, rather than the invention and refinement of theory?

Relationships with Views on "Scientific" Social Issues

Perhaps of most interest in discussions of STS education are measures of association between knowledge of science and position on social issues with scientific overtones.

Both the recent British studies made some assessments of the relationships between scientific knowledge and support for science. Durant and colleagues (1989, p. 14) do not present the data from their preliminary analyses, but do comment, "better informed respondents [tend to] have a more positive attitude towards science and scientists." This is consistent with the statistically significant pattern shown in Table 11.1, which comes

TABLE 11.1 *Support for science by science knowledge score*

Merits of science	Questions correct 0–7	8–15	16–24	Totals
Gone to far	94	252	21	367
Great benefits	76	376	75	527
Neither	19	56	8	83
No opinion	19	29	1	49
No answer	2	5	—	7
Totals	210	718	105	1033

$\chi^2 = 29.71, p < 0.001$ (Based on the 977 respondents who offered an opinion.)

from data collected in my 1986 study mentioned above. My question was: "Some people believe man has gone too far in tampering with nature through scientific enquiry. Others say this kind of scientific enquiry has brought great benefits to man and should be encouraged. Which comes closest to your view?"

Examining the relationship between scientific knowledge and position on a controversial issue is of more relevance than assessing generalized support for science. In my survey the questions about radioactivity and nuclear waste provided a simple measure of scientific knowledge on a specific issue, which could be related to beliefs about nuclear power policy. There was no significant relationship between giving an acceptable answer to the question about the nature of radioactivity and the respondent's opinion whether there should be more, less, or the same amount of electricity generated from nuclear power. Neither was there a relationship between the power-policy question and a question assessing the knowledge of the time that waste takes to decay. There *was* a relationship, however, between the power-policy question and the general science knowledge score ($\chi^2 = 13.42$, 4 d.f., $p < 0.01$), with the more knowledgeable tending to give greater support to increased power generation. But it is difficult to interpret relationships with total scientific knowledge scores because there are many intercorrelated variables. High scores are associated with (1) increased level of education, (2) interest in science news and TV programs *and* in most other hard news areas, and (3) socioeconomic status.

In summary, studies of the public understanding of science show many people unable to answer correctly simple questions of science. The public has at best a naive, and more realistically, a confused view of the epistemological foundations of scientific knowledge. Furthermore there is little relationship between specific scientific knowledge and preferred policy options in relevant areas.

However, before one uses these generalizations to urge action to correct "deficiencies," it is very important to put the generalizations in context.

A citizen cannot be expected to be equally attentive to all areas of knowledge or activity. Perhaps the scientific community, of which STS proponents are a subset, ought to be thankful that science commands as much interest as it does. Miller (1983) classed 18% of the U.S. population as members of the "attentive public." Moreover, 38% of the sample in the Durant and colleagues' (1989) study classed themselves as "very interested" (on a 3-point scale) in new scientific discoveries. In my study, 17% (on a 4-point scale) rated themselves as "very interested" in reading news of new scientific discoveries, and 22% said that they made a particular point of watching the television science series *Horizon*.

Attention and interest are not guarantees of high scores on general scientific knowledge. Less than 25% of the "very interested" group in my study answered correctly more than 15 of the 24 knowledge questions, although there *is* a statistically significant relationship between knowledge and interest, as shown in Table 11.2.

Studies of the scientific literacy of members of the public need to be tempered with the knowledge that members of specific social groups who have a utilitarian or hobby interest in a scientific field can become very knowledgeable about the science behind their concern. The motivation to develop their expertise may be political (for example, to help mount arguments against the "establishment" experts), or it may be related to a drive to have a greater understanding of a hobby interest. Recreational sailors can become expert in the interpretation and explanation of meteorological patterns, whereas other users of weather information may not seek to go beyond the short-term implications of the weather forecast (Lucas, 1985).

TABLE 11.2 *Relationship between scores on scientific knowledge test and interest in reading about news of new scientific developments*

	Number of questions correctly answered			
Interest level	0–7	8–15	16–24	Totals
Very interested	18	113	41	172
Fairly intersted	75	368	51	494
Not very interested	52	147	12	211
Not at all interested	62	85	1	148
No opinion	3	5	—	8
Totals	210	718	105	1033

$\chi^2 = 104.5, p < 0.001$ (Based on the 1025 respondents who offered an opinion.)

Layton, Davey, and Jenkins (1986) provide a full discussion, with examples, of groups that have special needs to develop their scientific literacy.

We must also remember that it is unrealistic to expect every member of the public to be equally knowledgeable about all areas of human intellectual life. What is the public knowledge of basic economic concepts? Of economic models? How many understand the nature of political processes? Furnham and Gunter (1987), for example, found that almost 70% of young people believed, incorrectly, that a general election must be held every 4 years in the United Kingdom, and that almost 60% believed, again incorrectly, that local councillors have the right to change laws which have been passed by Parliament. Even basic locational knowledge of geography is poor: "despite heavy U.S. involvement in the Persian Gulf, three-fourths of Americans cannot find it on a world map. Two-thirds cannot find Vietnam" (Grosvenor, 1988). Results like these are used by the relevant pressure groups to deplore the public's lack of knowledge about their subjects, and to urge that something be done about it through the school curriculum or via public education programs.

THE RESPONSE IN THE SCHOOL

As we have seen above, background knowledge, gained both from formal classes and from life experiences, is related to understanding science messages from informal sources. There is no reason to suppose that other areas of life are any different. (Examples of the use of background knowledge in museums can be found in Lucas, McManus, and Thomas, 1986; McManus, 1988; and Tulley, 1990.) We can expect people to interpret accounts of science-related controversies in terms of their understanding of the status of scientific assertions, and their knowledge of economics, history, psychology, and so on. With improved curricula, we can hope for a population that is at least aware of the fact that the science relevant to a policy issue does not determine the outcome, although perhaps it can set constraints. Policy issues are inevitably those where competing aims are judged, and the choices are determined by weighing consequences. If "the facts spoke for themselves" there would be no need for policy.

In the long run, STS advocates must join with other areas of the curriculum to produce an education in ethics and politics. To remain aloof and argue for a special role of science and technology in society (implying that the scientifically trained teacher is omnicompetent when an issue related in any way to science is discussed) is a form of the scientism otherwise rightly deplored by most STS activists. The most important danger is the possibility that science in the curriculum will obtain a stranglehold on these issues, thus denying the reality of the informal sources in which we are immersed. These sources, just like the issues, do not come in well-ordered, discipline-related packets.

Chapter 12

African Cultural Perspectives and the Teaching of Science

Olugbemiro Jegede

The characteristics of STS science education mentioned in Part I of this book account for why STS has spread to many parts of the world. The acceptance of STS by the developing world, especially the African continent, is enhanced by other factors related to the gaps, imbalances, and conflicts, that exist between a country's traditional orientation and the worldview of science and technology. As I noted a few years ago (Jegede, 1988), some of the most important reasons for the introduction of STS curricula in Nigeria (and indeed other African countries) include:

1. The general dissatisfaction with the ineffectiveness of the reformed science education curriculum in dealing with the social and other dimensions of science in relation to the immediate environment
2. The need to address the issue of the place of technology in a developing country
3. The need to harness the remains of the indigenous technology (for example, tanning, craft, painting, printing, and communication) with a view to integrating them with modern technology, requires a course like STS to provide a forum that involves the discussion of science and technology far beyond their facts and principles
4. Certain cultural beliefs and customs that the learner brings into the classroom are in opposition to, or incompatible with, modern science. (This leads to misconceptions, negative attitudes toward the study of science, and a confused psychological state of mind on the part of the learner.)

This chapter examines how non-Western sociocultural perspectives affect the teaching of Western science in Africa. In this context, STS education is seen as bridging the gap between traditional African values and Western scientific values because STS embraces the very sociocultural factors that seem to alienate the traditional African from Western science. I conclude with a proposal that would guide the development of STS education in Africa.

INTRODUCTION OF SCIENCE INTO AFRICA

There is evidence in the literature indicating that Western education was used, not to promote the healthy coexistence of the Western and the African cultures, but as a sanitizing and civilizing medium. As noted by Dart and Pradham (1976), "the attitude and often the intent of western education has been that a 'primitive' or 'decadent' civilization is to be replaced with a more 'modern' and 'better' one. . . . It tends to be particularly strong in science teaching, for science teaching is taken to be the one really unique and powerful offering of the western world" (p. 655). With the missionary urge to convert the "heathen," it has often been assumed that the so-called primitive races of Africa have no science (Maddock, 1981). The fact is that Africa had and still has its own science and technology. These were based on very different conceptual and cognitive models, which were not apparent to, or regarded by, the colonialists who first introduced Western education to Africa. The implications of this have been (1) the superimposition of one culture (Western) over another (African), and Western science over the indigenous science; (2) the attempt to completely wipe out the science and technology of the indigenous African peoples and hence their culture and history; and (3) the uncomfortable dissonance between the two worlds of science and culture that the learner is torn between in a non-Western environment. It is no wonder therefore that teaching science via the reformed science education of the 1960s, with its limited success in the Western societies from which it originated, was even less successful in a somewhat "foreign" environment.

NON-WESTERN CULTURES

To understand the problems that learners encounter in non-Western cultures when studying Western science, it is necessary to summarize some characteristics of non-Western cultures. Like most traditional non-Western cultures, African society is typically oral, with heavy reliance on interpersonal communication. Learning is a direct communal activity achieved mostly from a sage or elder. The learner is supposed to be passive and must defer to the supremacy and authority of the elder. African society is also characterized by a very strong social heteronomous organization.

Three fundamental features can be identified from the African belief system, especially the religious aspects of that system: (1) all Africans believe in the existence of the creator—the supreme God, (2) there is the belief in the continuation of life after death—reincarnation is a common and very popular aspect of the African culture, and (3) the human is seen and regarded as the center of the universe. The creator made human beings the focal point (Urevbu, 1987a). This anthropomorphic view of nature by Africans (Ogunniyi, 1988a) governs their thoughts, the way they do things, and the way they relate to one another within the community.

WESTERN SCIENCE AND AFRICAN CULTURE

On the one hand, Western science is mechanistic, exact, and hypothesis-driven. It seeks empirical laws, principles, generalizations, and theories. On the other hand, the African culture is "monistic-vitalistic" (Odhiambo, 1972), metaphysical, and based on people. Western science is seen, to some extent, as public property and divorced from religion. African cosmology is entirely secretive and interpreted as magical and is intrinsically intervowen with traditional religion. In the words of Mbiti (1969), "wherever the African is, there is his religion, he carries it to the fields where he is sowing seeds, to the beer parlour . . . to a funeral ceremony, to the examination room, if he is a politician to the house of parliament" (p. 2).

This means that, from the standpoint of cultural differences, different conceptual models of nature emerge reflecting systems based on prototype experiences (Horton, 1971). These two models are the mechanistic and the anthropomorphic, the former based on the Western scientific worldview, the latter based on the African traditional worldview. There is often more than one way to view an issue, if one tries hard enough. It is therefore natural for the African to view nature relative to the indigenous conceptual model. The problem, however, occurs when the African must learn Western science against the backdrop of myriad sociocultural factors brought from the traditional environment.

Although the study of science in most African countries is being vigorously pursued as a legitimate and effective means of national development, technological advancement, and scientific literacy, science teaching has remained largely restricted to giving information (facts) rather than encouraging critical and creative thinking and the personal construction of knowledge. In addition, the teaching of science in African classrooms does not emanate from, nor relate to, the sociocultural environment of the learner. It does not take into account, or seek to explain, daily personal experiences or interaction with technology in the rapidly growing and changing African society. This is where STS is of most worth and use in African classrooms.

SOCIOCULTURAL FACTORS

Culture has a great deal to do with differences in achievement in schoolwork (Jegede & Okebukola, 1988, 1989). Glasser (1991) asserts that cognitive activity (in school and outside) is inseparable from its cultural milieu. This has also been supported by anthropologists like Ogbu (1992), who say that school learning and performance are influenced by complex social, economic, historical, and cultural factors.

Because every society educates the younger generation as a means of passing down the sociocultural attributes of its people, the sociocultural factors within non-Western societies become a composite part of the environment. These factors control, to a very large extent, what a child in such an environment learns and what a child becomes in later life. Because every human "tends to resolve puzzles in terms of the meanings available in a particular sociocultural environment, the baseline is that the meanings become firmly implanted in the cognitive structure and manifest themselves habitually and may act as templates, anchors or inhibitions to new learning" (Ogunniyi, 1988b, p. 10).

Although efforts have been directed at unraveling other variables that affect science teaching and learning, very little effort is being directed at the sociocultural niche of our science classrooms. Studies in this area are needed for two main reasons. The first is global in nature. Culture, as the totality of all humans, subsumes every endeavour we undertake, including science education. Science education is a cultural and human enterprise involving the transmission of cultural heritage of a people (Gallagher & Dawson, 1984; Maddock 1981). Every investigation of human organization ought to, therefore, tackle the sociocultural issue. The second reason is personal in nature. It stems from my own experiences as a teacher of students in non-Western environments, as well as of students with traditional backgrounds in a Western environment.

Why is it that students in the two categories mentioned above show certain traits that do not seem congruent with what is expected of "normal" learners in science classes? For example, they hardly display the urge to ask questions in class. When forced to voice their opinions, some believe science has very little relationship with their own real world. Others think the study of science is a weird, special activity requiring some magical and superhuman explanations.

To the science teacher who perhaps shares the same sociocultural background as this group of students, the issue is as real as it is frustrating. The situation is even worse (and may be horrendous) for the teacher with a Western background who has to teach students of non-Western backgrounds.

There is an acute dearth of information to help teachers grapple with the effect of sociocultural aspects on the science classroom. Based on our personal experiences as science teachers and teacher educators who have had to address these aspects of the science classroom, we embarked on a series of studies in culture and science education (Jegede & Fraser 1989; Jegede & Okebukola, 1989, 1990, 1991, 1993; Okebukola & Jegede, 1990). The results of our investigations have led us to conclude that there are five predictors of sociocultural influences on the learning and teaching of science in Nigeria: authoritarianism, goal structure, traditional worldview, societal expectation, and sacredness of science. Each is described in turn. Their importance to science teachers cannot be overemphasized.

• *Authoritarianism* characterizes the traditional society where the belief is strongly held that the older person, having been exposed to more life experiences, should be in a better position to appraise a situation and pass "correct" judgement. The society frowns at a situation where the elder's point of view is challenged or questioned. Accordingly, the elder asserts authority in decision making. It behooves the younger individual to accept without question the directives passed down by the elder. This locus of authority of knowledge is transferred into the classroom where the science teacher is seen as the elder who "knows all" in matters relating to scientific facts, processes, principles, and laws. David Layton in Chapter 4 cautioned STS educators to consider this sociocultural factor.

• *Goal structure* refers to the interaction pattern among the people of Africa, a pattern that is predominantly cooperative in nature. In this cooperative setting, the goal structure of individuals is directed at the same objective, and there exists a high interdependence among the individuals working toward a goal. This contrasts very markedly with the individualistic competitive orientation that school science portrays to learners.

• A *traditional worldview* relates to traditional beliefs and superstitions being used as a framework through which occurrences are interpreted. African society holds the notion that supernatural forces do have significant roles to play in daily occurrences. The younger members of the traditional society are supposed to learn and believe these notions without questioning them. However, this creates conflict when the learner's scientific knowledge is not in agreement with the traditional worldview.

• *Societal expectations* play a pivotal role. The success of individuals within a community is developed and interpreted through the nature of their interaction within a communal society. The behavior of members in the community is invariably and intimately linked to, and governed by, the behavior of the larger community. Hence, a school child always reviews his or her achievement in school as a reflection on his or her home, friends, and community.

• *Sacredness of science* pertains to conceptual interpretations of science. There is a pervasive view held by a large proportion of African society that the study of science is something special in that it requires magical explanations incompatible with the thoughts of someone from a non-Western society.

There is reason to believe that the currency of some of these sociocultural factors may transcend international and regional boundaries particularly of Africa, Asia, and the Americas (Gallagher & Dawson, 1984; George & Glasgow, 1987). Understanding the sociocultural framework of a learner's mind is compatible with the emerging paradigm of alternative conceptions and constructivism in science teaching (Driver & Oldham, 1986). This will go a long way in helping to realize the five domains of science education as catalogued by McCormack and Yager (1989): knowing and understanding, exploring and discovering, imagining and creating, feeling and valuing, and using and applying.

STS SCIENCE EDUCATION

Science has become an international currency for national and global technological development that affects other sectors of human endeavor. Contemporary developments indicate that any nation that disregards this effect of science does so at its own peril. Indeed, the level of science and technology development in a country or region serves as an indicator of the general standard of living. Africa therefore cannot sit on the fence. The efforts that have been made regionally in Africa (for example, through the *African Primary Science Program,* APSP; the *Science Education Project in Africa,* SEPA; and within individual countries, such as the *Nigerian Integrated Science Project,* NISP) are positive indications that Africa is prepared to move with the trend.

However, this movement to transform the continent through science and technology has to grapple with a number of obstacles. These arise out of the sociocultural framework of the African traditional environment. For example:

1. There is a conflict of conceptual and instructional models between the school and the traditional society
2. There is an active interplay of psychosocial and sociocultural variables in the teaching and learning of science
3. The science learned at school has very little bearing on what goes on in the lives of individuals in a society that emphasizes human interactivity and communal living
4. The science taught at school consists mainly of facts, and places heavy reliance on rote memorization and regurgitation.

Given this situation, how can science be relied on as a viable avenue for development? These problems, although difficult to address, are not insurmountable.

I would like to suggest that embracing STS appears to be the most attractive viable option for Africa's educational system. This suggestion is based on: the international experiences of STS education over the past two decades, the limited but reasonably successful offerings of STS in some parts of Africa, and the potential it has for the present and future development of Africa. In addition, STS education should do the following:

1. Provide a driving force toward scientific literacy and science for all
2. Generate information about the African environment to explain natural phenomena
3. Identify and use the fundamental scientific and technological principles, theories, and concepts of the indigenous practices within African society
4. Arm the populace with vital information and explanations necessary to cope with the ever increasing demands of science and technology
5. Enable the people to develop indigenous technology, while appreciating the science behind it, and using their technology more extensively to solve human problems
6. Promote greater awareness and encourage the strengthening of the African communal spirit. (A communal spirit would be a central issue more than ever before as scientific and technological innovations penetrate the fabric of human relationships, potentially leading to extreme impersonality.)
7. Teach and uphold the value of the typical African humane feelings

in relation to, and in the practice of, technology as a human enterprise.

Some of the reasons supporting these suggestions are that STS science (1) brings scientists, science educators, and indeed all groups of people, to think of the scientific enterprise as a human enterprise; (2) relates what is taught in science classes to our day-to-day living; (3) is cognitively accessible at different levels to the majority of those who enjoy science studies, who want a science-based career, or who need it for matriculation or certification purposes; (4) uses local resources for the understanding of science concepts; and (5) demonstrates in concrete terms that science and technology are major factors that will influence the future of the whole world.

IMPLEMENTATION STRATEGIES

Restructuring our science education must be compatible with the objectives of science-technology-society (STS) education, and must be compatible with the African classroom environment described above by the five reasons for supporting STS science. Three major strategies for restructuring our science education are discussed here: (1) using familiar materials and processes, (2) teaching and learning science through indigenous technology, and (3) accommodation and assimilation.

Using Familiar Materials and Processes

Because science and technology are part of human culture, every society has some type of science and technology in one form or another. The way they are conceived, taught, and used, however, may be different in various cultures. For the African, the view of nature is quite different from views held elsewhere, for reasons already discussed. Western science embodies a language about nature spoken through the conceptual understanding of Western culture. Teaching Western science in the African society needs to be approached through materials and processes already existing in the traditional environment. Science education must begin from where the child is, even if it means beginning with the examination of the traditional belief system. After all, it has been established that Western science and superstition have common grounds (see Malinowski, 1948; Zar'our, 1972). As Zar'our asserted:

> Both superstition and science partially aim at explaining processes and suggesting schemes to predict events, interpret natural phenomena and to solve

> problems. Both assume implicitly or explicitly the ability of man to understand and to a lesser extent control his environment. (p. 279)

One way to help eradicate the fear or apprehension African children have towards science is to identify the elements of a number of fundamental scientific principles in some of Africa's so called fetish, primitive, or crude practices, and to link these practices to some Western science principles. Urevbu (1987a) suggested we begin to teach science by focusing on socially relevant issues that hinge on human activities. He cited an example of how Gay and Cole (1967) had taught science in Liberia by using as their starting point the mathematics of the people in everyday life. Biological, physical, and chemical principles relevant to some African beliefs, customs, and superstition are excellent ways of bringing home science concepts via familiar materials and processes. A few examples include: the blanching of the popular green leafy vegetable (Amaranthus), drum making and the effect of noise on humans, soap making, palm wine tapping, tie dyeing, and preserving local food stuffs. Empirical support for the need to use the cultural environment as a vehicle to teach science is found in the study by Jegede and Okebukola (1991). They investigated the relationship between African traditional cosmology and students' acquisition of a science process skill. In order to motivate understanding of scientific concepts and processes, science education for African children must begin with and reflect the worldviews that these children already possess, and then focus on issues that are relevant.

Teaching and Learning Science Through Indigenous Technology

Indigenous technology in Africa abounds in arts and crafts, hunting implements, pottery, food production, and irrigation, and so on. Teaching Western science can produce meaningful results when it is taught by using the African version of science and technology and by relating it to local traditions. Amara (1987) has found in Sierra Leone that using indigenous technology as a basis for science education encourages more female participation. In a culture where women suffer overt discrimination as byproducts of the sociocultural belief systems, Amara's findings will no doubt be a welcome development. Knamiller (1989) also linked school science with indigenous science and technology in Malawi and found:

> One of the reasons why school science has remained alien to most African children is because we often fail to take into account the science and technology local people are doing, what knowledge and skills they have and what problems they feel are important to consider. (p. 2)

Knamiller also found a common mission in a number of science projects going on in several African countries, including Zambia, Ghana, and Malawi. The aim of these projects was to package science and technology education materials meaningful not only for the minority of children who will go on to be scientists, engineers, and technicians, but also for the general mass of children for whom science and technology has a role to play in their private and civic lives. There are no better practical examples of STS education than the ones Knamiller has cited.

Accommodation and Assimilation

On the one hand, I am suggesting that the mismatch between science education and the African view of nature might lead to some cognitive conflict. On the other hand, I am advocating some form of accommodation and assimilation of the two views of nature. The prevailing science and technology in use all over the world originated in Western culture. Africa cannot and should not isolate its people from the influence of this Western science and technology. Africans need to understand the theories and generalizations behind some of the concepts and technological products they use daily.

We must resolve the paradox between (1) the African traditional societies in which people use technological materials but do not know any science related to those materials, and (2) the practices of indigenous science and technology that are not recognized by most Western cultures as valid. Drawing upon the theories of "psychic unity of mankind" and "cognitive restructuring," Ogunniyi (1988b) has shown that some form of "enlargement" of the African traditional worldview could be made to accommodate the Western scientific point of view. This, he notes, can ameliorate any possible conflict between the two systems of thought, as evidenced by the coexistence of Shinto and electronics in modern-day Japan. Accordingly, Ogunniyi has proposed the development of a science education curriculum that will not supplant or denigrate a traditional culture, but will help people meet modern challenges.

Urevbu (1987b) also subscribes to the idea that the African child can be helped to achieve a coexistence between cultural values and Western science. In recognizing that the purpose of education in Africa is *not* to destroy its own civilization in order to replace it with something conceived of as "better," Urevbu cautions that:

> An implacable either-or approach, leading to a direct confrontation between African traditional attitudes and a Western approach to knowledge, invites conflict both within the student's own mind and between students and elders

> in the community. Such conflict alienates the African child from one world without readily admitting him or her into another. (p. 291)

FUTURE DIRECTIONS

This chapter addressed the need to design science education that meets the needs of the African in such a way that the African view of nature, its sociocultural factors, and the logical dialectical reasoning embedded in African metaphysics, are established within a changing global community. Based on the evidence and arguments in this chapter, STS education would seem to be the most comprehensive, effective, and adequate way to achieve this.

Based on my personal experience and studies, and those of other African science educators, I propose that the best way to achieve the desired results is to use what I have called a *conceptual ecocultural paradigm*. It is a mechanism for designing science education that would prepare and propel the African continent into the next century. Conceptual ecocultural paradigm is a state in which the growth and development of an individual's perception of knowledge is drawn from the sociocultural environment in which the learner lives and operates. A conceptual ecocultural paradigm consists of:

1. Generating information about the African environment to explain natural phenomena
2. Identifying and using indigenous scientific and technological principles, theories, and concepts within the African society
3. Teaching the values of the typical African humane feelings in relation to, and in the practice of, technology as a human enterprise.

The paradigm addresses two major, interrelated, educational issues that have emerged recently. The first is the issue of constructivism, in which the learner constructs his or her own knowledge from new experiences and in terms of an existing conceptual framework. The second issue relates to the worldview that learners in traditional societies take into the science classroom. The learner's worldview acts as a framework within which the mechanistic science concepts are assimilated.

Science educators are aware of the imperative to close the gulf between the learner's sociocultural environment and school science. Africa needs STS education as a vehicle to reach this desired goal.

Chapter 13

Innovative Science Teaching in Rural Schools of India: Questioning Social Beliefs and Superstitions

Anita Rampal

Soon after the monsoons in 1991, a particularly potent rumor gripped large parts of the country, causing tremendous apprehension and fear. The story spread like fire, especially through the rural countryside. It told of the widespread appearance of curious white snake-like lines on green leaves, and invoked the divine "curse" of a deadly "serpent god." People stopped eating vegetables afflicted by the "snake-lines." Cartloads of gourd were thrown into rivers. Mounds of coconuts were offered to appease the *naagdevtaa* (the serpent god), while women with divergent religious persuasions observed weeklong fasts. Media reports amusedly recounted regional variations on the serpent story, but failed to give any explanations for the phenomenon. A member of the Upper House questioned the government for not having taken appropriate steps to disabuse people of the belief that had made them shun essential vegetables. Subsequently, the state-run television station broadcast a program for farmers in which experts ritually pontificated on the harmlessness of the afflicted leaves, but they were ineffective in demystifying the issue.

Within this milieu of uncertainty and anxiety, the response of a few young school girls commands attention. They dared to discover the real culprit for themselves, and even managed to challenge the community belief. The girls achieved this feat quite unselfconsciously because they had been part of an inquiry-based science program that trained them to perform simple experiments on their own. They had not studied in an elite private school endowed with fancy equipment, but instead, in a very ordinary government school housed in characteristically dingy and cramped quarters. Furthermore, these were girls from ordinary working-class backgrounds whose parents, themselves uneducated, were already planning to marry them off at the young age of 14, as is customary in that society. Their story is taken up later in this chapter.

THE HOSHANGABAD SCIENCE TEACHING PROGRAM

Science in India continues to be taught in a didactic and conservative chalk-and-talk manner, with only a few experiments ritually conducted by students in the final years of high school. In addition, many students never make it to high school because an emphasis on rote memorization of remote concepts in a formidable foreign language alienates the majority of young children, and they drop out long before they complete elementary school. The Hoshangabad Science Teaching Program, HSTP (Mukund, 1988; Rampal, 1991, 1992b), made a significant break with tradition when it developed a curriculum with a sense of "criticality," while meaningfully addressing the life experiences of predominantly rural children. The program recognized that openness of thought needed to be complemented by an appropriate degree of socialization that allowed for free inquiry and democratic interaction both inside and outside the school. Many issues needed to be addressed, including conservative attitudes towards: the curriculum, examinations, girls and other socially marginalized children, and the teacher in the community and in the classroom. The image of science was also mystifying and alienating.

HSTP was initiated as early as 1972 by two volunteer organizations engaged in local developmental work. Academic support was provided by university scientists who, themselves inspired by the curricular reforms in the West, willingly took time off from their regular assignments to work in the program.

The traditional curriculum continues to be controlled by the demands of university science. Its abstract academic nature supports the minuscule percentage of school children who may eventually become professional scientists (Rampal, 1992a). HSTP, on the other hand, supports the STS movement by giving priority to fundamental skills and concepts that would help students to develop a critical scientific attitude for life. Students who follow the traditional curriculum suffer the additional hegemony of a scientific discourse in a "sanskritized" or classicized form, in the absence of a popularly acceptable lexicon. HSTP text materials, however, are more colloquial in nature, using simple everyday words as much as possible.

HSTP alleviates the sense of alienation normally associated with science lessons by using low-cost equipment, indigenously designed apparatus, and locally available materials. Moreover, the attempt to do away with unnecessary trappings that had to be bought from far-off places symbolized an assertion of self-reliance and provided an incentive for teachers to exercise their own creative skills. For instance, many of the earlier ready-made bulb holders and cell holders were replaced by simple substitutes devised by teachers. The HSTP began an almost revolutionary principle in our country by demanding that all elementary school pupils handle apparatus fearlessly, and by compelling the government to provide for normal breakage.

HSTP rejected the usual practice of pushing new curricula onto unsuspecting teachers, who tend to follow their own instincts of practicality rather than the optimistic idiom of planners (Calderhead, 1982; Doyle & Ponder, 1977). Instead, the HSTP considered school teachers as the protagonists of educational reform and collaborators in curriculum development. The distinctly democratic culture of friendly discussion and fearless inquiry provoked teachers to "de-expertize" the urban experts so that the program could emerge as a truly shared construction. Expert drafts of chapters or ingenuous designs of experiments were willingly discarded when teachers expressed reservations or children failed to respond satisfactorily. Passive acceptance of subject authority, or of any authority for that matter, was to be almost assiduously rejected.

Teachers were intensively trained to conduct activities and discussions in the classroom. These new methods were often a difficult experience for the teachers (Buseri, 1987; Olson, 1982). Notwithstanding the initial resistance to change, a large proportion of teachers responded enthusiastically and put in tremendous extra effort to make up for their own lack of training in the subject and to satisfy the ever-increasing demands of their excited pupils. HSTP believed that if progressive reform were to be successful, it would not be through special schools and selected superhuman teachers, but through the government structures (however dilapidated, leaky, and ill-equipped) and through the ordinary teachers (after they had been suitably galvanized to usher in the desired change).

HSTP has since expanded to over 500 schools, spread over many districts of Madhya Pradesh (an educationally and economically backward state in central India). It is administered by the state department of education, though the academic responsibility is still borne by the voluntary group "Eklavya." During these 20 years, HSTP has inspired many groups elsewhere to work toward curricular reform, and has been hailed as a major innovation in science education in the country.

MOVING TOWARD A SCIENTIFIC TEMPERAMENT

The refrain of "progress through the inculcation of scientific temperament" has reverberated across the Third World during the last few decades, but there are problems. An overly positivist and often erroneous view of science was seen as the panacea for all national problems. Secondly, scientific thinking was naively perceived to be in direct confrontation with people's religious and customary beliefs, superstitions, and traditional practices.

Despite our constitutional commitment towards scientific temperament, not much has been achieved and no effective methodology evolved either by the state, which controls both education and the media, or by the different political parties working at the grassroots level. In fact, the right-wing parties have systematically invoked religious dogma and superstition to incite communal passions, with alarming repercussions in the last few years. Meanwhile, standard courses in school science, which allow no room for thought and reflection and impart no skill to deal intelligently with everyday objects, have failed miserably. So has the television media, which mindlessly beams a barrage of often crude and condescending messages, with no attempt to communicate explanations. For example, the media patronizingly orders viewers to wash their hands before they eat, or repeatedly reminds them to vaccinate their children. Moreover, the messages seem to place the blame entirely on the poor viewers for their illnesses, almost absolving the state of its responsibility to provide the most basic amenities, such as potable drinking water. Addressing people's beliefs and superstitions is clearly a complex and delicate business requiring a hermeneutic approach with sufficient time, understanding, patience, and empathy to present an alternative worldview without causing offense.

HSTP was mindful of the traditional religiosity of most teachers and local resource persons, and was cautious in its approach so as not to alienate or belittle them. Science was seen as a specific system of knowledge, based on empirical validation, which did not necessarily interfere with one's personal philosophy or mode of worship. Such issues were discussed openly, not uncomfortably brushed aside. Teachers were generally curious to know about the views of scientists and asked questions about the conceptions of God held by Einstein or Raman. Interestingly, many teachers tended to regard scientists stereotypically as worshippers of nature (not religious in the manifest sense of the term) whose humility before the "sublime force" is taken as a natural substitute for a more ritualistic faith (Rampal, 1992c). Discussions regarding philosophical positions on agnosticism, the limits of scientific knowledge, and other topics have also been conducted. Many of us were wary that science also runs the risk of being placed on an artificial pedestal, itself taking the position of an uncontested faith.

Superstitions and taboos were distinguished by the teachers from their religious beliefs. They were ready to accost their own inhibitions, after mentally preparing themselves for the change. It was indeed heartening to see how even older teachers, who had long resisted touching an egg or who had thought they could never bring themselves to dissect a frog, gradually overcame their reservations and became the most active supporters of this method of teaching. Both men and women freely sought scientific information about normally taboo topics such as menstruation and human reproduction. They readily accepted alternative explanations to commonly held beliefs. The pupils of these teachers naturally gained some confidence in dealing with local beliefs, as was spontaneously demonstrated by many of them in the dramatic incident of the "cursed leaves."

CHILDREN'S RESPONSE

The most unreserved response to the changed climate of the HSTP classroom came from children, who thoroughly enjoyed the new curriculum. Science lessons became conspicuously animated and noisier, in almost cathartic defiance of the traditional norms of "pin-drop" silence. The high point of excitement was the excursion into the fields to collect samples of soils, plants, or insects. Orthodox parents of girls initially objected to this unconventional freedom, but were soon reassured by the teachers regarding the safety of their wards as well as the necessity of the "excursion" chapters.

Rural children are generally familiar with the natural world. HSTP offered them a systematic method to conduct simple observations and experiments. They tested and dispelled many commonly held beliefs, such as those about spontaneous generation, flies from cowdung, or frogs from rain water. In some schools, children have persisted for more than a month, fascinatedly observing the "birth" of a frog from the larvae they collected, even changing the water every few days by fetching it from the original pond one to two kilometers away! Indeed they have often gone beyond the textbook, excitedly collecting eggs of lizards, lice, and cockroaches, to see their little ones emerge.

Girls have generally shown maturity and responsibility in handling equipment, unlike boys that age who seem to need more policing. Teachers have often entrusted girls with full charge of the kit. The changed role from passive receiver to active participant has helped in building the self-confidence and critical attitude of these girls, who often tend to be deprived of such opportunities at home. In fact, the more active girls have often become upset at home with the refrain that these science lessons have made them "too curious" and "too defiant"!

MEETING THE INVESTIGATORS OF THE "SERPENT" STORY

The young girls who exposed the "serpent" rumor had also earned a similar reputation for being curious and defiant in their homes. "We have experimentally disproved many superstitious notions. Despite being told not to, we deliberately walk past when a cat crosses our path or when someone sneezes, and have found that nothing happens. People sometimes reprimand us and taunt us for having been spoilt by our education."

These girls were frank, forthright, and surprisingly self-confident. They said they had heard the serpent episode authoritatively narrated by reliable sources. They believed the part of the story about the truck driver who was responsible for crushing the male snake having been cursed by the female consort. However, what the girls found incredible was that the curse could result in snakes "appearing" on all leaves. They cursorily dismissed this popular explanation with "snakes are too big to become manifest on a leaf," or "it cannot happen to millions of leaves simultaneously!" They said they suspected it was due to a small creature "which leaves behind its saliva," about which they had learned in school.

They raised the issue in class to check on what their teacher believed, and on finding her equally inclined towards a possible physical rather than a supernatural cause, were further encouraged in their quest. Two of them went home and inspected the affected leaves in the backyard, and after attempting to dissect a large number of them with a pin, finally succeeded in exposing the tiny culprit, less than a millimeter in size, embedded deep within the leaf. The next day the entire school was deftly hunting for this minute "worm" and excitedly examining it through the simple plastic microscope provided in the HSTP kit.

This was a major feat, for even some scientists who had elsewhere been queried on this phenomenon, and were convinced it was caused by either a virus or an insect, had taken some time to finally locate the culprit, after having duly checked the literature for relevant information. As it turned out, this was apparently a larvae that buried itself in the leaf and fed on the chlorophyll, thus leaving a colorless trace behind. The worm seemed to grow bigger as it ate, causing its trace to gradually widen and appear more like a serpent with a thin tail and a wide head on either side.

FAITH AND BLIND FAITH

"These ideas are related more to blind faith and are less worthy of belief," was the profound sentiment expressed by one of the girls. In her attempt to change her family's beliefs, she, who normally did the cooking at home, had even tricked them into eating a dish made from the affected tomato plant, challenging them the next morning to show what harm had occurred! On being asked about the possibility of being thrashed for this she laughed, saying, "One has to be prepared for that in order to remove blind faith." Encouraged by their confident articulation, I ventured to probe somewhat deeper. If this was blind faith then what was it they considered worthy of faith? They went on to explain how they believed in telepathy (without being able to name it). "When my heart suddenly beats faster, I know that something is wrong at home." They also said they were wary of experimenting with "religious" beliefs, such as not looking directly at the devi (the goddess) during the 9-day festival of prayers, and felt threatened by the elders' insistence that their irreverence could cause the death of someone close to them. Curiously, the curse of the serpent-god had no religious origin for these children, even though it invoked a similar threat of death, while on the other hand, beliefs associated with the family deity held greater sanctity.

This interaction revealed deeper philosophical underpinnings of the role of faith in cognition. It also indicated how the categories of superstition, religious belief, and local taboo are socially and culturally constructed. Each community asserted its set of beliefs, exhibiting varying degrees of faith towards each family of such beliefs. However, when ready to question certain ideas, as this young generation had been adequately socialized to do, even a fragile experiment, however disputably designed, seemed to provide sufficient refutation to scaffold the process of disbelief.

LOOSENING OF "CLOSED" SYSTEMS

In the Indian context, it is difficult to distinguish among religious, magical, and traditional beliefs. They occupy complementary domains in the space of social cognition. Such beliefs broadly constitute what Horton (1970) calls a "closed" system of thought, characterized by a lack of awareness of alternatives to the established body of theoretical tenets, and constrained by the tendency to reason only within a "limited idiom." Moving toward a scientific way of thinking would involve developing an awareness of possible alternatives, and allowing traditional tenets to become less sacred as they lose their absolute validity, accompanied by a loosening of social structures that sustain such tenets.

The demarcation between science and traditional knowledge is not sharply perceived in India. For instance, indigenous systems of medicine that have a professed empirico-rational basis (such as Ayurveda, Unani, or homeopathy) are popularly regarded as "scientific," as distinct from the other prevalent folk systems founded on more magico-religious practices. Similarly, most people do not differentiate between astrology and astronomy, and are unable to see why astrology is not science. In fact, commercial astrologers, who were amongst the first users of the computer in India, benefitted from the false scientificity the computer lent to people's implicit faith in their profession. People are thus easily confounded on two accounts: they view the computer as a legitimizing symbol of modern science, and they tend to misconstrue historical accounts of the early Indian contributions to astronomical science as being credible corroboration of their faith in astrology.

HSTP has tried to analyze the earliest elements of scientific practice in prehistoric and present-day tribal societies. However, this narrow path needs to be tread cautiously. On the one side lie the right-wing chauvinists, only too keen to eulogize and glorify past achievements of Indian science and now dangerously involved in perpetuating their own myths. On the other side are the antiscience propagandists, equally ready to misinterpret any critical view of established science.

HSTP has stood for informed criticality in trying to understand science, its limitations, and its distinction from nonscience. However, HSTP's thrust is reflected more in its teacher trainings, informal publications, and regular activities for children than in its text materials, which have to conform to the state curriculum.

Some of us now hope to be able to radically restructure the texts in the light of our consolidated STS experience, as well as to compile STS readers for teachers. HSTP has encouraged a better understanding and appreciation of traditional empirical systems of knowledge possessed by farmers, artisans, and housewives. HSTP has tried to project the method of science as accessible to all, rather than as a body of unquestionable expert knowledge.

Part IV

Gender Questions and STS

Any movement committed to making science relevant to all people cannot ignore half the population. When women already feel alienated from the existing "valid and neutral" science, as described in Part I, their omission would be a scandal. Two chapters in this section offer a broad sweep of gender questions.

Chapter 14 by Joan Solomon begins by recalling an early British link between STS and the liberation of women. Then the deficit view of girls in science education is described. The deficit approach is based on the simplistic interpretation of test results showing that girls do poorly, or at least less well, than boys. She recounts how the more perceptive research methods of the late 1980s were needed to exhibit the positive characteristics that girls bring to the learning of science. Finally, where some in England seem to see no way forward except by teaching girls on their own, sequestered from the possibly overbearing influence of boys, Joan Solomon's work envisages quite a different solution. She uses data from a research project about how boys and girls discuss STS issues in the classroom to show that the complementary attitudes that girls and boys often adopt can benefit the whole class.

The second chapter in Part IV has quite a different perspective on gender questions. Written by a leading feminist scholar in Britain, it traces the progress of the radical critique of a masculine science from the points of view of capitalism, paternalist domination, and the new environmental movement. Hilary Rose probes the "telos," or final purpose, of science and technology institutions. This leads to a consideration of the position of women within science and science laboratories. Her main thesis is that a new epistemology of science can be drawn from her radical perspective on history, sociology, and women's awareness. This is of great importance for the philosophical and sociological strands in STS thinking. Her synthesis of the different modes of human activity—hand, brain, and heart—also speaks powerfully to STS educators who strive to forge ways of teaching science that are more truly an education for the whole child, girl or boy, than what has gone before.

Chapter 14

Learning STS and Judgments in the Classroom: Do Boys and Girls Differ?

Joan Solomon

The aim of this chapter is to chart how the ship of STS sailed through the troubled waters of the gender question in science. The voyage went from a deficit model of under-achieving females, through an exploration of gender differences of attitude and behavior, to a stage where research could explore how boys and girls in STS lessons profit together from new and appropriate teaching strategies.

It is not possible to encompass all the gender-related work that has been carried out during the last 20 years. What is described here comes mostly from Britain. The reason for this lies partly in my greater knowledge of such work in Britain and partly in the existence (up to 1989) of student choice whether or not to study science at school. This made the preferences of the genders a matter of particular importance.

EARLY *STS* AND THE LIBERATION OF WOMEN

The gender question arose in the public consciousness at almost exactly the same time as the STS movement began. In Britain, an organization of tertiary science teachers from universities and polytechnics led by Bill Williams, then of Leeds University, initiated the SISCON (Science in a Social CONtext) project in the early 1970s. It was dedicated to exchanging ideas and producing new teaching materials that could extend the learning of science from theoretical abstractions to the real problems and concerns of contemporary society. This was a very diverse group and, in the beginning, many of its members were radical thinkers. Those who led SISCON were associated with a number of avant-garde movements so, in this sense at least, it was not surprising to see them also backing women's opportunities and using their limited funds to second a graduate student to write one of the earliest of all teaching units to be entirely devoted to women in science. In those days in Britain it was still more difficult for women to enter medical school than it was for men, and there were very few women in scientific posts in the universities.

There were stronger reasons than simple inequality of opportunity, however, for including gender questions in STS. The final initial in "STS" meant that society needed to be empowered to use the knowledge about science that had recently given rise to the "superstar technologies" of the 1960s. That "s" meant all of society, not just the male 50%. But there was an even more important point. SISCON-type courses studied the conflicting interests of different sectors of society; thus it was entirely understandable that women's interests should require separate deliberation. At SISCON meetings with Dutch STS colleagues, special all-women's meetings were scheduled as early as 1975.

I have the impression that the first consideration of gender within the STS movement was simply modernist. It aimed for fairness and equality of opportunity. The school version of SISCON (begun in 1978) was led by women science teachers working in public comprehensive schools in inner-city areas where inequality of many sorts—poverty and race as well as gender—was everywhere apparent. In many British schools, girls routinely studied only biology, which was itself something of an advance on the times. Only a decade earlier, "boys took science and girls took domestic science."

THE DEFICIT MODEL

In the 1970s and 1980s, it was normal to begin a book or article about gender in science education with statements like the following: "This then is the intellectual problem: why do girls, relative to boys, underachieve in science?. . . Girls who succeed in science have a wider choice than those who fail, so the feminist seeks to reduce failure" (Kelly, 1981, p. 13). Girls' failure was the problem to be tackled. Such sentiments seemed to be entirely unexceptionable. Reliable statistics from many countries were collected then, as they are today, to show that females do less well than males on school science tests (for example, Comber & Keeves, 1973; Erickson & Erickson, 1984; Johnson & Murphy, 1986). Attitude tests also showed girls to be less positive towards science and technology, then as now (Breakwell, 1990). There can be little doubt that these data are "correct" or, at the very least, reproducible. The result is a clear deficit model of the female failing in science. The deficit model won very wide credence and led to some measure of remedial action.

In the last century, skull size or biological function would have been suggested to explain the poor science performance of females. In the middle of this century, the results of tests on "spacial ability" were given in evidence, later to be examined and rejected by Macoby and Jacklin (1975) and Harding (1983). Later still, the causes of failure were thought to be: social stereotyping, textbook illustrations, no opportunity to tinker with toys, exclusion from apparatus by boys, or lack of role models. The fact that the explanations for the poor showing of girls in traditional science classes were moving away from phrenological measurements to other areas was cause for only mild satisfaction.

The early 1980s saw a large-scale action research project, *Girls into Science and Technology* (GIST), which took place in some 20 British schools (Whyte, 1986). The main strategy was to combat the public image of scientists as exclusively male. A smaller but similar attempt to "feminize" the image of science was attempted in the United States (Kahle, 1987). Schools in the GIST project used textbooks with females in the illustrations and arranged for female scientists to visit the schools and talk with students. These schools also included stories about female scientists in their teaching. And yet the boys' and girls' attitudes towards science, although affected, remained almost as far apart as ever.

The total evidence collected showed only that girls' performance was poorer and that girls professed less interest in science or technology in all attitude measures. Attempts were made to ensure that teachers paid full attention to the girls in the class, and to finding out what aspects of science content interested girls most. These interests did not match the boys'. Girls were different, it seemed, in all respects. Looking back it is all too easy to be wise in hindsight by insisting that the characteristics of girls' approach to science had not been explored in any depth at all nor had any attempt been made to explain them. The term "girl-friendly," however, with its rather patronizing overtones, was to be heard everywhere.

Yet a few authors, such as Harding (1983) and Smail (1984), had drawn attention to teachers' observations that girls seemed interested in the social issues (or was it just the social applications?) of science.

RESEARCH INTO GENDER CHARACTERISTICS IN SCHOOL SCIENCE

Within the next 5 years, there was to be some excellent work done on the psychological aspects of gender. Within science education, Head (1985) pursued the theme through published accounts of several studies showing that many of the British boys who so often elected to study physics did so more for reasons of lack of maturity than of intrinsic interest. School physics seemed to provide them with a reassuringly male and authoritarian route where difficult reflection on sensitive personal matters was unlikely to occur. This was the first indication that conventional school science, especially the physics by which girls seemed most repelled, may have been fashioned to a schema that was emotionally skewed. For any students who wanted to discuss controversial social issues, traditional school physics was to be avoided like the plague. Some boys and many girls were doing exactly that.

In 1990 Head and Ramsden reported a series of research findings of interest to STS educators. They found that the established and expected connection between psychological type, gender, and science subject choice completely broke down when "the physics topics were contextualized so as to emphasize the social and applied importance of the work" (p. 119). Nor was this simply a quantitative effect. Head and Ramsden pointed to a whole new psychological population type being attracted to this new social approach to physics.

There is little space here to do justice to the other studies that slowly began to overthrow the deficit model in favor of one that tried to explore what it was that interested girls, and why fewer of them enjoyed traditional science. In 1984 Collins and Smithers published the results of a very large questionnaire study that included batteries of wide-ranging questions. This classic paper, "Person Orientation and Science Choice," had originally been stimulated by the psychological observations of adult scientists. Their research concluded that "Person orientation emerged as an important discriminator in every analysis undertaken and outweighed all other personality constructs in the various multi-variate statistical analyses" (p. 62).

In the context of STS studies, two other complementary research findings of this period need to be mentioned, however briefly. Both relate to girls' unwillingness to answer questions with a simplistic "yes" or "no." In Likert tests (Grant, 1987) and in multiple-choice examination questions (Murphy, 1990), girls showed a stronger preference than boys for considering the uncertainties, or describing both sides, of the question, and were therefore often reluctant to opt for brief positive or negative responses.

STS EDUCATION: HOW DO GIRLS BENEFIT FROM IT?

This rather partisan series of abstracts, from a substantial body of research into girls in science, points in one direction. Many girls tend to be most interested in topics that relate to people and their predicaments. Where adolescent boys sometimes shy away from consideration of socially complex issues requiring empathic thinking and leading to no "right answer," many girls are attracted to such study. The implications of these findings (in terms of the choice of course content and teaching strategies) clearly point toward the implementation of a full STS program of study. In the examination-led science curricula for 16-year-old students in British high schools, this was hard to achieve in the early 1980s.

One little known study stands out, however. It investigated the success of girls in "advanced level" physics courses, the area of most concern to teachers and industrialists (although for widely different reasons). Pell (1985) conducted a longitudinal survey of more than 700 students in 16 schools over a period of more than 5 years. Pell's work examined various measures of enjoyment and achievement. He was able to show that at 16 years of age the enjoyment score was by far the best predictor of examination success, especially for girls. As students progressed into the "advanced" physics course, the girls who had chosen this course actually maintained their interest better than the boys, some of whom may have chosen it for reasons of prestige rather than intrinsic satisfaction. For the boys, finding the subject "easy" correlated highly with enjoying it. For the girls, the relationship was much more complex. Factors such as "considering social implications," "thinking deeply about your personal views," and "[taking] a more philosophical approach" all contributed to the girls' enjoyment of physics and to their level of achievement.

Research on gender within British STS courses such as SISCON was rare, probably because such novel educational ventures originating from teacher-led movements attracted minimal funding. The gender differences that turned up during small-scale action research were often striking and unexpected (Solomon, 1985). At first these differences were also hard to understand. The fact that the issue was "social" was not enough to give any indication of what approach was required. In my 1985 study, a paper-and-pencil exercise was carried out with over one hundred 14-year-old physics students. The students' sentences about alternative energy strategies were found to be gender dependent. The girls tended to write with more concern and urgency. Figure 14.1 exemplifies the writing of girls and boys.

Although these findings suggest that the girls were more emotionally

Girls	*Boys*
We must look for other energy sources.	Alternative kinds of energy are being looked into.
They have to be found.	Scientists are looking for...
(I think) nuclear energy may be dangerous.	Methods of disposing of nuclear waste include...
People are worried about...	They are researching...

FIGURE 14.1 *Typical sentences about alternative energy strategies*

involved in the work, and although other findings showed that they did well on such courses (Pell, 1985), the research had other implications. STS teaching cannot escape from moral issues (see Chapter 11) and it began to seem as if the girls welcomed talking about these issues more than the boys did.

MORAL JUDGMENTS: KOHLBERG AND THE FEMINISTS' RESPONSE

It is well known that Kohlberg and his colleagues examined the stages in childhood moral development in a way that was consciously fashioned in the Piagetian mold (Kohlberg, 1984). In his original work, Kohlberg followed a sample of boys over many years as they progressed through a series of stages from "being a good boy" to learning to "care for other individuals," to a reflective understanding of disembodied notions such as "justice" and "human rights." Kohlberg was later to find that females did not fit at all well into his scheme. They seemed to remain, for the most part, obstinately at the third stage, arguing in the context of care for individual people rather than at the higher abstract level. For Kohlberg and his coworkers, the dilemma was either that women had to be considered incapable of full moral development (the deficit model again), or that they had been retarded by being kept at home where their moral worth could only be exercised and demonstrated in this inferior way (caring for people).

The argument was taken up by feminists such as Carol Gilligan (1983). In general, these writers made no attempt to contradict the findings that women perceived moral dilemmas differently from men. Gilligan herself made two parallel distinctions. The first was about the role of self for males and females respectively: "the contrast between a self defined as separation, and a self delineated through connection, between a self measured against an abstract ideal of perfection and a self assessed against particular activities of care" (p. 35).

A very similar point about the sense of connectedness (girls) or separation (boys) was made by Wingfield and Haste (1987) on the basis of their empirical interview work on moral dilemmas. The second contrast that Gilligan made (perfection versus care) arose from work carried out in a preabortion clinic and refers back to the Kohlbergian theory of moral development. She did not deny difference between the genders, but saw female adolescents' moral progress not toward abstractions, but toward a resolution of people's conflicts. She wrote that women see "moral dilemmas in terms of conflicting responsibilities" (p. 105). This position was later confirmed by Gibbs, Arnold, and Burkhart (1984).

When teaching science through STS, issues arise quite naturally in the classroom, issues that affect people's well-being and question conventional notions of responsibility. Issues such as pollution by industrial employers, and energy obtained at the expense of risk to a local minority, need to be discussed between girls and boys with their science teacher. By 1990, the social context of science was accepted to the extent that the Policy Document on Gender and Science Education, written by the influential Association for Science Education, included on their list of required "Approaches to Teaching and Learning":

> (f) provide opportunities for pupils to make evaluative statements about the role of science and technology in society, and
> (g) develop the "knowledge and understanding" objectives in the context of social and personal responsibility and morality. (Association for Science Education, 1990, p. 2)

DISCUSSION OF ISSUES IN THE CLASSROOM

Feminist research, as we have seen, had emphasized the differences between female and male approaches to social and moral debate. There was clearly an empirical question about how these might affect learning strategies in mixed-gender classrooms. The DISS project—the *Discussion of Issues in School Science* (Solomon, 1992c)—provided data on how boys and girls discussed issues in STS lessons. It set out to explore how 17-year-old students argue about science-based social issues communicated by means of general release television programs that students have been watching together in the classroom. The work ran in 12 different schools during 2 years. We used classes of students taking a course in STS science. On a series of occasions during the school year, one of six video excerpts started the discussions on controversial topics. The teachers did not ask any specific questions, but simply encouraged the groups of three or four students to talk together about the issues, as they saw them, and to record their discussion on audiotape. A total of 217 discussions of various lengths were recorded and transcribed.

The origins and characteristics of the students' knowledge were examined by pre- and postcourse questionnaires. Another set of data was provided by a profile of the characteristics that each student displayed during discussions. The questionnaires were quite substantial, containing sections included for comparison with those from other projects in the program. The section most germane here asked about the nature and origins of the students' views and knowledge about science-based issues. The gender differences in their responses are presented in Figure 14.2 as a cluster of

Girls	*Boys*
When certain I try to convince others.	Those who worry about issues are probably ignorant.
Most of my knowledge about issues comes from TV.	My science teachers do not mention issues
My views on issues are probably like my friends'.	Issues are best left to the scientists
More likely to watch TV than go out with friends.	My views on issues are probably like my friends'.
My views on issues are probably like my parents.	Issues mentioned are all good for society.

FIGURE 14.2 *Main pre-test factors*

questions that factor analysis showed to be responsible for most of the variance in girls' and boys' responses.

In ways that are not unfamiliar, we can see in Figure 14.2 that girls tend to present themselves as more likely than boys (1) to have opinions that have been socially constructed by collaborating with friends, and (2) to have watched television. The boys tend to accept scientific outcomes much more uncritically. The work of Head (1985) and Pell (1985) had suggested such results.

From the discussion transcripts, 12 characteristics were selected. The students were scored for each occasion in which they showed:

- Knowledge from television or radio
- Demand that more research be done on a topic
- Vague need for action—"They should . . ."
- Frustration at lack of available knowledge
- Comment for or against industry
- Intention to respond by individual action
- Idea for a civic strategy
- Knowledge from reading
- Persistence in argument—"Yes, but . . ."
- Collaboration in sentence construction or repetition of phrases
- Attribution of bias in knowledge presented
- Confidence in giving out new items of knowledge.

To a greater or lesser extent all of these are characteristics of the social construction of knowledge. Some relate to action derived from knowledge, and some to behavior in the social reception of knowledge. Independent analyses of the transcripts were carried out by two observers, who scored each student who had taken part in at least four discussions (only 145) on each of the 12 characteristics above. There was better than 90% agreement among the observers. The scores were then amalgamated and entered as "none," "only once or twice," or as "more than twice," to form a discussion profile for each of the students.

Analysis of these amalgamated scores showed surprisingly little gender difference. Only in the case of verbal collaboration did we find that the girls scored higher than the boys (significant at the 5% level). This finding is consistent with the notion of "connectedness" mentioned in the previous section. In terms of giving out items of knowledge (whether right or wrong!), the boys scored slightly higher than the girls (significant at the 10% level). On all the rest of the profile items, including knowledge from television and persistence in argument, the girls did not score differently from the boys despite their different presentation of themselves by questionnaire responses.

CATEGORIZING MORAL STATEMENTS

The work of Kohlberg and his critics suggested that careful note should be taken in the DISS project of the ways in which the two sexes addressed questions of social justice. At first this presented great difficulties. Different topics produced very different types of moral response; for example, individualistic on kidney donation, political on nuclear power, and sex-related on genetic counseling. In the end, one topic out of the six was chosen—the case of leukemia among the veterans from the Christmas Islands tests of the atomic bomb. Three categories of moral statement were created. Each is considered in turn.

Broad Statements

"Broad statements" have some of the characteristics of the higher levels in Kohlberg's description of moral reasoning. They comment on the problem or issue from a decontextualized position; for example, "It's the principle not the money."

The majority of the students' broad judgemental statements, in which they stood back from the details of the situation, focused more specifically on what the government should do, the servicemen's rights, or the regulations that should be enforced. "They should be allowed compensation because it was not voluntary." "Yeah, I mean that isn't right. You shouldn't have to pay with your life."

Arguments concerning scientific or mathematical thinking were also placed in this category; for example, statements where the speakers used numbers to explore the validity of the disputed statistics. This is one of the logical skills that Piaget associated with formal thinking in his stage theory of cognitive development. Because Kohlberg was much influenced by Piaget, as noted above, abstract thinking in mathematics sits comfortably in this category.

Personal Statements

At the opposite extreme is a type of comment that shows the speakers so strongly empathizing with the people in the video that they try to include themselves, or their friends, in the situation. "I'd rather die first if they dropped the bomb anyway—I'd go and sit myself underneath it." "Would *you* worry?" The speakers attempt to understand the issues by imagining how *they* would feel, and what *they* would do if directly involved. Kohlberg placed such statements ("putting yourself into other people's shoes") down at stage 3 in moral development.

Invented speech put into the mouths of others was quite common as a rhetorical device intended to persuade. But it also served the same purpose as the statements quoted above in making the situation seem more immediate and thus easier to comprehend. "It's like the government said, 'You are 18 years old. We want you to join the army for 6 months'." "It's not as though they said, 'Oh I *want* to go down to be 4 or 5 miles away from the bomb'."

Contextual Statements

With "contextual statements," the speakers take account of the situations of the people and events in the video, and add comments based on their own experience or understanding of similar situations. Some showed strong emotive reactions, others were cooler and even cynical, and many simply recalled factual details from the video that seemed relevant to the present stage of the discussion. For example, "The children are the ones that are going to live with it." "I think it could be because [the official] is probably in favor of the government, and he could be a bit biased."

Conclusions

Using a double-blind examination of unmarked scripts, a high degree of reliability on the categorization of statements (broad, personal, contextual) was achieved, with better than 95% agreement. Some general conclusions can be drawn, most of which do *not* show gender differences.

1. By far the largest number of contributions from both girls and boys were contextual. The students seemed to be using their talk to get a grasp of the situations and needed to mull over the details of the story in order to give their impressions of it and underline some of the issues.
2. Both boys and girls made broad statements during discussion more in accordance with individual or group styles of argument than with gender.
3. The opening statements of the discussions served managerial purposes. In one way or another the students (girls or boys) seemed to be attempting to get the discussion going. There were several tapes showing a hesitant start with words like "Right . . ." or "Well . . ." Others began almost formally with "We're talking about test veterans . . ." (See also Barnes and Todd, 1977.)
4. Personal statements were *not* made more often by girls than by boys in the main sequences of discussion.

Further Analysis

The opening statements of all the discussions were reexamined separately from the main discussion. The rationale for this was based on the observation that only at the start of a discussion would gender differences stand out clearly without the blurring mediation of social influence. Although the topics under consideration varied, and this undoubtedly affected the nature of the main sequences of discussion, all topics could supply useful evidence about opening statements. Marked gender differences surfaced immediately among the 91 opening comments (Solomon & Harrison, 1991). The boys were more than three times more likely to start the discussion with a broad statement than were the girls. No such gender differences were found for either contextual or personal statements. The results are shown in Figure 14.3.

Judging by the satisfaction expressed by the students themselves, one of their main aims in the discussions was to arrive at an agreement about the actions that should be taken to redress wrongs. Statements in the main body of the discussion often included reflections on civic strategies for dealing with the issue. These were about legal constraints or procedures that were thought to guard against future risk, or to compensate for an injustice. For example, "[T]here should be some laws. . . ." "You can't put a price on that . . . they should put [the money] into a big fund which you take out as you need it." It was striking that in no case did these strategic and civic statements follow a broad comment used to open the discussion. This may be taken to confirm that these openers were recognized as triggers for argu-

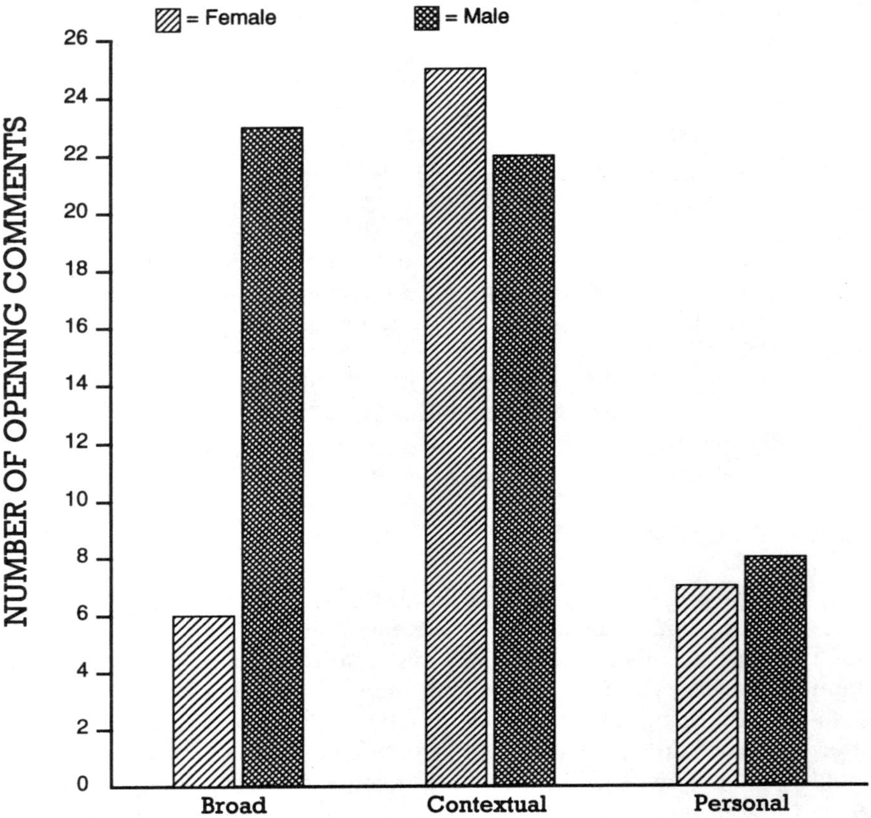

FIGURE 14.3 *A comparison by gender of three types of statements that initiated student discussions*

ment, rather than considered judgments. It may also suggest that the striking gender differences shown in opening sentences were accepted as matters of preferred style.

SOCIAL EFFECTS AND GENDER

The DISS project research took place in STS science lessons. There were some gender differences, but these were far less than those listed by many earlier researchers. Wingfield and Haste (1987) had placed so much emphasis on the gendered use of empathy that the personal statements made by students during the discussions in the DISS project were reexamined with the greatest care. Personal statements seemed to serve at least three purposes: rhetorical mocking by imagined role-play questioning to get another's opinion, and imagining what it would be like if they were in the situation. Even on these sub-categories no significant gender differences could be found.

Contextual talk might also have been expected to show gender differences because it figures prominently within the feminist literature. Keller's (1985) analysis of the work of the Nobel Laureate Barbara McClintock referred frequently to "getting a feel for the context," a phrase that distinguished McClintock's work from that of male scientists. Once again, however, there was no difference between the girls and boys during mixed group discussions.

To understand the DISS gender findings it is important to remember that these were the only data to have been drawn from the students' behavior in a social situation. It may well be that the nature of social discourse has itself served to modify gender imbalances. The work of Moscovici (1976) and others has indicated the influence that individuals can have within groups. The purpose of social talk is to try out opinions, receive feedback, and respond to others. This necessitates taking on board others' perspectives and even their style of discourse. Thus the immediate social influences (the questions and comments of peers) may well override gender-related styles and orientations.

The study by Wingfield and Haste (1987), which found such convincing numerical evidence for gender effects, was based on interviews of individuals. It is interesting that the only similar gender differences to be found in group discussions were in the opening statements, where social pressures from others had not yet begun to take effect. However, Wingfield and Haste did note that each gender can reproduce the style of the other. In the DISS project discussions, because of their social nature, styles of talk needed to be continually varied and matched to the comments and questions of the other students. An inability to respond in terms of the others' relevancies would have indicated a severe lack of social skill.

The deficit model of girls in traditional science and technology classes has led to some calls for segregated teaching. Such an outcome might well be considered a failure by STS teachers who aim to provide young people with an education in which they consider the concerns of society at large. The boys and girls taking part in the DISS project spoke intently together, using their knowledge of science and society to exchange ideas about governmental responsibility and social justice. Their suggestions for social strategies were built on the understanding of a range of opinions and moral positions.

The deficit model of the female in science has no relevance for these discussions any more than it has for the social or philosophical aspects of STS. Science-based social issues have proved to be accessible to both girls and boys, in single sex or mixed sex groupings. It is only when these aspects of science are omitted from lessons that more girls than boys have been shown to find little interest in the subject. The opportunity to learn about and discuss social issues gives all students the sort of citizenship skills that could be practiced throughout life, and to which STS education is so strongly committed.

Chapter 15

The Two-Way Street: Reforming Science Education and Transforming Masculine Science

Hilary Rose

Being a scientist or engineer in the West is much like being a member of any other relatively well-paid and interesting occupation, that is, it is largely the preserve of white and able-bodied men. However, although most people interested in science and science education would concede this generalization, a number would feel uncomfortable with its bleakness and would want to point out the efforts that have been made in recent years, particularly by science educators, to broaden and diversify the pattern of recruitment. Others point to the present difficult climate for science educators. Sir David Weatherall, clinical geneticist and current president of the British Association for the Advancement of Science, has drawn attention to a pervasive anti-science stance, particularly among the young, whether female or male. Although I do not think Sir David's concept of anti-science does justice to the complex social criticism made against science by the young and the not so young, what his remark does is to acknowledge a link between the discussion of the future scientific labor force and the question of the content of science (Rose, in press).

To give an everyday example, numbers of young people are committed to animal rights. In consequence, some refuse to carry out dissections as part of biology. Such a refusal may not be part of a generic anti-science stance, but may be a very specific rejection of what geneticist J. B. S. Haldane spoke of in his later years as "violent biology." Such "dissection refuseniks" might well be very interested in the nonviolent biology that Haldane sought to foster. They may not declare themselves to be environmentalists or feminists, but they are located in a sort of cultural soup in which these powerful intellectual currents eddy around.

This chapter argues that we cannot separate the discussion of science education from the question of who are to be the producers of scientific knowledge, or from the question of what kind of science gets produced. Thus, in the previous chapter, Joan Solomon addresses science education as an issue where gender is central. In this chapter, I address the masculine nature of science and suggest how a concept of responsible rationality could foster the development of sciences akin to Haldane's notion of a nonviolent biology. Both preoccupations reflect a view that at the close of the 20th century, science and technology are much too important to be left exclusively, or even mainly, to one gender.

Nonetheless, the political objective of getting more women into science has not necessarily been shaped solely by desires for gender justice or by dreams of a new and more responsible science. The one time that women are welcomed into science is during war, when suddenly unmoveable barriers can be dissolved. For that matter, the surrogate war of competition in space, notably Sputnik in the fifties, also saw the barriers suddenly soften. It is only under certain conditions that governments look to women as an undertapped supply of scientific talent. What has been distinctive about the last two decades, and their intensifying pressure for changing the gender composition of the scientific labor force, has been as much a result of pressure from below, from the women's movement. Nowhere has this pressure been more effective than in the United States, its relative success evidenced in the statistical data produced by numerous governmental agencies monitoring the gender changes in science education and in employment.

Although it is perhaps a little heretical to say so in a book on science education, my suspicion is that if girls and women saw that employers, whether academic or industrial, were actively willing to hire and promote women scientists and engineers, and to put in the necessary social supports to make their participation practicable, the message that science was welcoming to women would reach down into the process of science education, making the task of transforming the pedagogy easier. The difficulty for science educators is that girls and boys can read the sharply gendered message still being sent from the scientific labor market, including that of science education itself.

But if changing the labor market and science education is now only beginning, the other and more complex task of changing the content of science is even less well developed. The first hints of such a change are, however, being made by the U.S. and Canadian governments. There, biomedicine's practice of studying the male body as a stand in for that abstraction, the human body, has been challenged, and projects that should appropriately include females, are, as a condition of federal grant support, now required to do so. In Britain and Europe, the customary practice reigns supreme except where an individual researcher successfully resists the inbuilt androcentrity. If this generalization appears gross, think of heart disease—where there is unquestionable evidence that the pattern of women's morbidity and mortality is quite different from that of men, and, equally unquestionably, that the bulk of heart research studies men. Biomedicine is only now (under pressure from the feminist movement, and with some reluctance) accepting the heart-related disease and death of women as a research issue.

The feminist call for the transformation of science, and with it science education (to embrace an STS perspective), is in tune with these postmodern times in that it is simultaneously both piecemeal, as in the very specific demand that science also studies the hearts of women, and revolutionary, in that it calls for nothing less than a new science. It is to this piecemeal revolutionary call that I now turn.

This chapter starts from the position that the cultural values of science and technology must be transformed, for their telos is that of global pollution and nuclear annihilation. However, changing the culture of science and technology is peculiarly difficult because they claim to have no culture, to be outside society, independent, and value-free. New social movements of this past quarter century, notably the radical science movement of the sixties and seventies, and more recently those of environmentalism and feminism, have laid political and intellectual seige to this claim. From within the academy, a parallel move by the new social studies of science has taken place that also points to the social shaping of science and technology. This new approach includes historians, philosophers, and sociologists of science; their foundational text is unquestionably Thomas Kuhn's (1970) *The Structure of Scientific Revolutions,* first published in 1962, which acted as both a landmark and a catalyst.

The chapter begins by examining the achievements made during the 1970s by the radical science movement as it sought to analyze and critique capitalist science's existing forms and systems of knowledge. I then argue that their critiques (which sometimes were developed with a conscious opposition to sexism in everyday life) were theoretically sex blind. Thus their analysis of the division of labor stopped short at the distinction between the manual and mental labor associated with production. Indifferent to the second system of production—reproduction—the analysis excludes the relationship of science to patriarchy, to the sexual division of labor in which caring labor is primarily allocated to women in both unpaid and paid work. Transcendence of the divisions of labor between hand, brain, and heart makes possible a new rationality. In its turn, this new socially and environmentally responsible rationality opens the possibility of a new scientific knowledge and technology that will enable humanity to live nonviolently with nature, including human nature.

THE NEED FOR A NEW SCIENCE

During the seventies, the critique of science and technology focused attention on the ways in which existing science and technology are locked into capitalism and imperialism as a system of domination. This denunciation has served two functions. Negatively, it has facilitated the growth of an antipathy to science that rejects all scientific investigation carried out under any conditions and at any historical time. More positively, it has set itself the difficult task of constructing, in a prefigurative way, both the forms and the content of a different, alternative science—one that anticipates the science and technology possible in a new society and, at the same time, contributes through innovatory practice to the realization of that society. This chapter aligns itself with such a venture, while recognizing that (with its false starts as well as real achievements; its perilous balancing between atheoretical activism and abstract theoreticism) the project is not without its contradictions and difficulties. Feminism is just beginning to recapture the full force of Virginia Woolf's (1952) insight: science, it would seem (to rephrase her), is neither sexless nor classless; she is a man, bourgeois, and infected too.

The trouble with science and technology, from a feminist perspective, is that they are integral not only to a system of capitalist domination but also to one of patriarchal domination; to try to discuss science under both these systems of domination is peculiarly difficult. Historically, it has been women outside science, such as the novelist and essayist Virginia Woolf, or the exscientist, now writer, Ruth Wallsgrove (1980), or the sociologist Liliane Stéhelin (1976) who have dared to speak of science as male, as part of a phallocentric culture. For women inside science, protest has been much more difficult. Their numbers are few and developing the network among isolated women is intractable work. Yet, as we enter the nineties, an invisible college of feminist scientists has clearly assembled. One of the early voices breaking through from within the laboratory is that of Rita Arditti (1979) who, having worked in the competitive and macho world of genetics, was radicalized by the antiwar movement. She became a feminist through this experience and now argues that nothing less than a new science will serve.

Apart from a handful of feminists who wrote pioneering papers, in the early days of the movement feminists avoided the discussion of science, often retreating into a total rejection of science as the monolithic enemy. Doing science became an activity in which no serious feminist would engage. But there arose a growing political threat as, during the 1970s, a new wave of biological determinism sought to renaturalize women. It has required women biologists (or women who will enter the terrain of biological knowledge) to contest its claims. Doing biology was no longer seen as hostile but as helpful to women's interests, and increasingly it has been possible to go forward from this essentially defensive purpose to the much more positive goal of seeking to show how a feminist knowledge of the natural world offers an emancipatory rather than an exterminatory science. The task of developing a feminist critique of existing science, and of moving toward an as yet unrealized feminist natural science, is at once more difficult and more exciting than the academically respectable activity of making descriptive reports on women's position within science or on the problems of girls learning science. There is a watershed between work carried out within a transformative view of women's destiny and projects shaped by the main variants of structural functionalism still extant. For feminist analysis, unlike developed Marxist theory (though like early Marxism), calls for interpretation constantly tested not simply against the demands of theory, but always and incessantly against the experience of women and the attempt to overcome women's oppression.

THE RADICAL CRITIQUE OF SCIENCE

The critique of science was to explode into practice and to struggle into theory during the radical movement of the late 1960s and early 1970s. The numerous issues contained in the class and social struggles of the movement were frequently narrowed and constrained as the theoreticians filtered the wealth of lived experience through the abstract categories of theory. From an early rhetoric that attacked with a certain evenhandedness the class society, imperialism, racism, and sexism (those who were black, colonized, or women might well have had doubts about their equal prioritization in practice as well as in rhetoric), the theoreticians were to develop two main lines of analysis. The first considered the political economy of science, and the second took up the relationship between science and ideology. Although the two are linked at many points, work in political economy was more coherently developed; work on the debate over science and ideology was and remains more problematic. The need to reply immediately to the ideological attacks of a racist or sexist science accelerated as the crisis deepened; it was difficult simultaneously to resist the attack and to analyze the issues. Indeed, the hostility to science within the movement so conflated science and ideology that it aided the growth of attitudes that rejected science altogether (Rose, 1979).

The socialist tradition, at least up to the sixties, believed that the advances of science would automatically create problems that capitalist society could not solve; hence, in some way science was at least "neutral," at best allied to those working for a new and socially just society. Such a neutral science was seen as uninfluenced by class, race, gender, nationality, or politics; it was the abstract accumulation of knowledge (of facts, theories, and techniques) that could be "used" or "abused" by society. The experiences of the sixties and seventies overthrew such notions of science. What the sixties' radicals discovered in their campaigns against an abused, militarized, and polluting science was that those in charge of neutral science were overwhelmingly white, male, and privileged occupants of positions in advanced industrialized society. The antihuman technologies that science generated were being used for the profit of some and the distress of many. Thus, the politics of experience brought the radical movement's attitudes toward science into a confrontation with the orthodox Marxist analysis of science.

THE MYTH OF THE NEUTRALITY OF SCIENCE

Although I have initially focused on the writing of those who were influential within the radical science movement, over the two decades or so there has also been a dramatic shift in the history, philosophy, and sociology of science. A sophisticated form of externalism holding the thesis that scientific knowledge is structured through its social genesis has become common to all three, so that research now focuses on the social construction of scientific knowledge (for example, Fee, 1980; Haraway, 1978). Thus, although Thomas Kuhn's work marked the beginnings of the thaw for an age that seemed forever frozen in the timeless certainties of positivism and the Vienna circle, it was a mathematician, historian of science, and political radical, Jerome Ravetz (1971), who posed the question of why and how science is a social problem. To answer this, Ravetz examined the circumstances in which scientists actually produce scientific knowledge. Abandoning the heady and very abstract Popperian theory of "bold conjectures and refutation" (Popper, 1963), he asked: What do scientists actually do? Through an examination of the production of science from the 17th to the 19th century he argued that, whereas in the early period science was considered a craft, it increasingly adopted industrialized methods of production as it entered the 20th century. Where the craft worker had labored alone, or with a couple of apprentices, the new system required substantial capital, a large group of scientists, a clear and hierarchical division of labor between them, and common goals to be set and managed by a scientific director.

Ravetz held that this industrialization of science has produced its uncritical character. In an essentially romantic and libertarian political response, he called for the deinstitutionalization of science. When tied to the state and industry, science must inevitably lose its critical force and become an agent of oppression. If a true science is to reachieve the liberating role it had in the time, say, of Galileo, it must, like Galileo, once more stand in opposition to institutionalized knowledge. It must become critical. The problem with Ravetz's position is that it is idealistic. Although a handful of deinstitutionalized scientific ventures came into existence, it would be unrealistic to consider their contribution a sign of the restructuring of science because their access to the means of scientific production is minimal.

Less attracted than Ravetz by alternatives outside science, and more oriented toward contesting existing science, others within the radical science movement were nonetheless pursuing the same theoretical concerns. They were revolted by the genocidal science that the United States employed in the war in Southeast Asia and by the expanding new technologies of urban repression at home. They asked, How can science claim to be ideologically pure, value-free, and above all neutral, when even a well-regarded text entitled *The Scientific Method* offers as an example of scientific development the making and testing of napalm on a university playing field, without any references to ethical or political problems? From the "use and abuse" model, in which science remained fundamental, basic, and pure, though possibly abused by political others, the New Left (to the equal concern of both the scientific establishment and the Old Left) had laid siege to the myth of the neutrality of science itself (Rose & Rose, 1979).

The change in the mode of scientific production, its loss of criticality, and its subjugation to the laws of commodity production, are features of the sciences most closely integrated with the reproduction of social and economic power. The physical sciences, above all physics itself, are at once the most arcane and the most deeply implicated in the capitalist system of domination. At the same time, the industrialized sciences more or less successfully exclude any more than small numbers of women. They also appear to be highly resistant to feminist reconceptualization; the success of feminist theory has lain in areas such as history, philosophy, sociology, and primatology—all characterized by little capital equipment per worker and by craft methods of production.

THE SECOND SYSTEM OF DOMINATION

Despite the advances made through the critique of science pursued during the 1970s, the critique is, in a theoretical sense, sex blind. It is not that the critics have been insensitive to the problems of sexism and racism; many have honorable records in trying to contest them. It is rather that the theoretical categories make it impossible for them to explain why science is not only bourgeois but male. For it is unequivocally clear that the elite of science—its managers and the constructors of its ideology—are men. Within science, as within all other aspects of production, women occupy subordinate positions, and the exceptional women who make it in this man's world only prove the rule. Yet this exclusion of half of humanity means that the 1970s' critics of science, while they grappled with the structuring of science and technology under capitalism, failed to grasp the significance of (or even to recognize) the structuring under patriarchy, that other pervasive system of domination. Indeed, the radical science movement itself was to reflect in its practice much of the sexism of the social order it opposed. Nonetheless, the critique laid successful siege to claims that science and technology transcend history, and made plain as well the class character of science within a capitalist social formation. Science as an abstraction came to be analyzed as an ideology having a specific historical development within the making of capitalism. Demystifying science exposed the myths that had served to integrate science and mask its internal contradictions and external functions.

Yet, within the radical critique of science, there remained a disjuncture between actual struggle, on one hand, and theorizing, on the other. Looking back over the writing of the sixties and early seventies, it is difficult not to feel that, as the critical work became more theoretical, more fully elaborated, so women's and men's interests receded. Thus, this writing gave no systematic explanation of the gender division of labor within science, nor, despite its denunciation of scientific sexism, did it explain why science so often works to benefit men. By attributing the exclusion of women to ideology, it ignored the possibility that there is a materialist explanation, nowhere hazarding the suggestion that it is in the interests of men to subordinate women within, as well as without, the production system of science. It is taken for granted that the domination-subordination, oppressor-oppressed relationship between men and women is either irrelevant or is an artifact of the production process.

WOMEN SCIENTISTS IN THE MEN'S LABORATORIES

Women who manage to get jobs in science have to handle a peculiar contradiction between the demands on them as caring laborers, as the mothers of children, as wives, and as the daughters of aging parents, and also as abstract mental laborers. Many resolve this by withdrawing or letting themselves be excluded from science; others become essentially honorary men, denying that being a woman creates any problems at all. This denial is particularly evident in the autobiographical accounts of successful women in the sciences. It took Anne Sayre's passionate defense of Rosalind Franklin to demystify this sexlessness and to insist that the woman scientist is always working in the men's laboratory (Sayre, 1975).

Evelyn Fox Keller (1977), writing of her experiences as a student physicist and later as a research worker, echoes this theme: the continuous, subtle, and not-so-subtle exclusion mechanisms deployed against women scientists. She writes that as a student she had to be careful to enter a lecture room with or after other students; if she entered first and sat down, men students found it threatening to sit near this low-status person—a woman student—and she was often surrounded by a "sea of seats." On one occasion, when she solved a mathematical problem, the male university teacher was so incredulous that Keller, like Naomi Weisstein (1977) in a similar situation, was quite gently asked who (which man) did it for her, or where she got (stole) the solution. Keller's experiences are not, however, unique; what is new is that they are discussed.

Thus a woman scientist is cut in two. Her involvement with the abstraction of scientific practice as it has developed under capitalism and patriarchy is in painful contradiction with her caring labor. As Ruth Wallsgrove (1980, p. 237) writes, "A woman, especially if she has any ambition or education, receives two kinds of messages: the kind that tells her what it is to be a successful person; and the kind that tells her what it is to be a 'real' woman." Small wonder that women, let alone feminists, working in natural science and engineering are rarities. It is difficult enough to suppress half of oneself to pursue knowledge of the natural world as a woman; it is even more difficult to work as a feminist scientist (Roberts, 1981). Part of working as a feminist involves the construction of an epistemology where thinking from women's everyday lives opposes the abstraction of male and bourgeois scientific thought.

RECONCEPTUALIZING SCIENCE

Feminist theorizing about science is of a piece with feminist theoretical production. Unlike the alienated abstract knowledge of science, feminist methodology seeks to bring together subjective and objective ways of knowing the world. It begins with and constantly returns to the subjective shared experience of oppression. It is important to stress *shared* experience because the purely personal account of oppression, while casting some brilliant insights, may tell us more about the essentially idiosyncratic character of individual experience than about the general experience of all or even most women. Nonetheless, within feminist theoretical production, experience, the living participating "I," is seen as a dimension that must be included in an adequate analysis. The very fact that women are, by and large, shut out of the production system of scientific knowledge, with its ideological power to define what is and what is not objective knowledge, paradoxically has offered feminists a fresh page on which to write. Elizabeth Fee (1981) pursues parallel themes. Largely ignored by the oppressors and their systems of knowledge, feminists have necessarily theorized from practice and returned theory to practice.

Although it would be false to suggest that all work claiming to be feminist achieves this dialectical synthesis, there is a sense in which theoretical writing looks and must look to the women's movement rather than to the male academy. Working from the experience of the specific oppression of women fuses the personal, the social, and the biological. It is not surprising that, within the natural sciences, it has been in biology and medicine that feminists have sought to defend women's interests and advance feminist interpretations. To take an example: menstruation, which so many women experience as distressing or at best uncomfortable, has generated a tremendous amount of collective discussion, study, and writing. A preeminent characteristic of these investigations lies in their fusing of subjective and objective knowledge in such a way as to make new knowledge. Cartesian dualism, biological determinism, and social constructionism fade when faced with the necessity of integrating and interpreting the personal experience of bleeding, pain, and tension (Hartsock, 1983).

Many of the slogans as well as titles of books and pamphlets arising from the movement speak to this necessary fusion. "A woman's right to choose" has, whatever feminism's difficulties with liberal individualistic theory, made immediate everyday sense to women. It has articulated the desire of women to recover the control over their own bodies, a control that the male-dominated medical professions and the profit motive have all too commonly appropriated. Self-examination and self-health care groups did not only offer prefigurative social forms of health care, but also prefigurative forms of knowledge. Rewriting biomedical science, women sought to reconstruct women's nature. The best selling and widely translated *Our Bodies, Ourselves,* produced by the Boston Women's Health Collective in 1971, is an icon of this intellectual and political process. The collective sought to reclaim a woman's sense of being a whole person living in society and not simply an assemblage of body parts. In a similar vein, *For Her Own Good* (Ehrenreich & English, 1978) not only affirms woman's capacities to understand her interests, but also exposes the masculinist professionalization of medicine in which alienated forms of both knowledge and care have driven out nonalienated feminine forms. In this situation, a feminist biology does not attempt to be objective and external to the female biological entity; it attempts to make over biological knowledge in order to overcome women's alienation from our own bodies and our own changing selves.

Nor was the feminist reconstruction of biology limited to the biomedical; feminist primatologists and paleoanthropologists critically examined the origins story of humanity because it historically represented women as inferior and weaker. Their work led to the representations of patriarchal hunting societies gradually being replaced by more compelling accounts of hunter-gatherer societies, in which both women and men contributed to the survival and well-being of the group. The account of the centrality and dominance of male primates similarly ceded to a story of female primates actively doing child care and managing to forage simultaneously. The success of the feminist primatologists' account supports feminism's political project of valorizing women's everyday lives (Haraway, 1990).

The creative energy of the women's movement in simultaneously fashioning new organizational forms and new knowledge was initially almost taken for granted by the movement itself. In its early phase, it made fresh syntheses of theoretical significance for reconceptualizing the natural as well as the social sciences with the disarming charm of Moliere's hero who discovered that he had been speaking prose all his life. This has given way to a situation in which feminists are exploring the epistemological transformations of both the natural and the social sciences; although there is little agreement, there is a growing confidence that what feminists have to say is important (Rose, in press).

The formidable and oppressive powers of science and technology, both ideologically and materially, continue to grow, whether evidenced as the eugenic undertones of the Human Genome program, as pollution from Chernobyl, or as the genocidal firepower displayed in the Gulf War. My argument has been that these are not some accidental consequence of science (some unfortunate use or abuse), but stem directly from science and technology's location within society. In consequence, the need for a new, nonoppressive science and technology is acute. While some environmentalists argue for a deep ecology to transform science from a knowledge of domination of nature into one of cooperation, I see the intimate connections between the construction of science and the construction of Western masculinity as impeding that welcome development. Instead, I see feminism, or rather the feminisms (as there are many, not one), as key agents in changing science and technology, along with other intellectual and social movements for change.

Where the socialist, critical thought of the seventies explored the division of mental and manual labor and its implications for alienated knowledge in the production of things, in the eighties feminism pointed to the third and hidden division of caring labor in the alienated reproduction of human beings themselves. In the nineties feminists speak of "thinking from women's lives" (Harding, 1991). It remains that bringing this hitherto excluded thinking into education about the natural world becomes critical for a transformative program equally within science, technology, and society. An STS education constructed from the practices of the hand, brain, and the heart offers, conceptually and practically, to deal more gently with the natural and social worlds.

PART V

RESEARCH INTO STS EDUCATION

In the previous parts of this book, the authors referred to research to warrant their claims about STS education. Part V gives emphasis to the research itself.

In Chapter 16, Glen Aikenhead reviews an extensive research literature to shed light on whether students actually benefit from studying science through an STS approach. The research quite consistently points to positive outcomes of STS education, especially for projects having a research and development (R & D) focus. This finding lends credence to STS projects and programs described throughout this book.

While Chapter 16 reviews the results of research, Joan Solomon in Chapter 17 critiques the research itself, uncovering difficulties that challenge all researchers, for instance: (1) the need to explicate how learning achievements come about, rather than to be satisfied with reporting significant differences between an STS group and its control; (2) the need for a more flexible conception of a constructivist's view on learning; (3) the serious flaws in research into students' attitudes; and (4) the problematic relationship between attitudes and subsequent social action.

Chapter 18 reports on research with teachers who were required to implement an STS syllabus. Peter Fensham and Deborah Corrigan employed an exemplary interpretive research model (strongly endorsed by Joan Solomon in Chapter 17) to investigate the ways in which teachers and course writers shared the same meaning of STS content and STS instruction. The in-depth longitudinal design allowed the authors to compare what teachers thought a year before, and just after, they taught the STS course.

Harrie Eijkelhof's exemplary R & D project (Chapter 19) demonstrates a formative evaluation strategy and a quantitative way to decide what knowledge is of worth in STS science education. His research, in part, systematizes the choice of science and technology content to be taught in an STS module on ionizing radiation, sharing with Herb Thier and Barbara Nagle (Chapter 8) a commitment to using content from the everyday world of experts who deal with the knowledge and with the public on a day-to-day basis.

A different type of R & D project, one inspired by the interpretive genre of research, is described by Glen Aikenhead in the last chapter. His work drew upon student experiences and natural classroom events to validate both course content and teaching methods in the development of an STS textbook for high schools. As a result, his materials are richly detailed in student contributions, and his research uncovered complex interactions that begin to explicate how students learn STS science, a challenge posed by Joan Solomon in Chapter 17.

Chapter 16

Consequences to Learning Science Through STS: A Research Perspective

Glen Aikenhead

Rhetoric, good intentions, and enthusiasm aside, do students actually benefit from studying science through an STS approach? Science educators have claimed that there is insufficient research on STS instruction to answer the question with confidence (Cheek, 1992a; Eijkelhof & Lijnse, 1988; Finley, Lawrenz, & Heller, 1992; Rubba, 1987b). This chapter comes to a different conclusion: Students can benefit from learning science through STS, and they benefit quite consistently.

Arguments in favor of STS education often assert lofty goals; for instance, to produce wise decision makers, an empowered citizenry, an enlightened democracy, a socially responsible corps of scientists and engineers, or an increased number of scientists and engineers (particularly women). These goals involve incredibly complex psychological and sociological interactions. Although such goals may inspire educators, their complexity defies evaluation. Lofty goals, therefore, make poor criteria for evaluating student outcomes of STS instruction.

More modest aims of STS education will better serve as criteria. Most STS science syllabi embrace four common aims (Aikenhead, 1986; Bybee, 1985b; Eijkelhof, 1990; Solomon, 1993): (1) to increase citizen's scientific literacy; (2) to generate student interest in science and technology; (3) to encourage interest in the interactions among science, technology, and society; and (4) to help students become better at critical thinking, logical reasoning, creative problem solving, and especially decision making. (See Part I and Chapter 5 of this book for an extensive description of STS aims.) Different educational groups around the world will operationalize these common aims differently (for example, Bybee, 1987; Chambers & Turnbull, 1989; Cheek, 1992c; Fensham, 1988; Waks, 1987b; Yager, 1992a).

The diversity of viewpoints on STS informs this chapter's review of research on student outcomes of STS science instruction. Different researchers will investigate different outcomes of STS instruction because of the researchers' different conceptions of STS (ranging, for example, from case studies in the history of science to political action on a local controversy). This chapter does not presuppose a single definition of STS for all to use. Instead, it accepts the diversity of views on STS that have guided research studies in science education. The chapter also examines an often-heard fear that STS science may not adequately prepare students for future science classes.

A two-part question, therefore, frames this research review: Do students actually benefit from, or are they harmed by, studying science through an STS approach?

The scope of my review is necessarily delimited. These delimitations are identified in the next section in a way that sheds light on the research review itself.

QUALIFICATIONS TO THE RESEARCH REVIEW ON *STS* INSTRUCTION

Science education research of the 1960s reached an unambiguous conclusion: the classroom teacher will influence student outcomes far more than specific curricula, textbooks, or teaching strategies (Welch, 1969). Thus, student outcomes from the same STS course can vary significantly from one teacher to another. Within any population of teachers there will be three groups: (1) those whose philosophy of science education is consistent with an STS approach, (2) those who are diametrically opposed to an STS approach, and (3) those in the middle who can move in either direction due to persuasion or by the requirement to use certain materials. All three groups will have their own influence on student outcomes. The research data presented in this chapter relate to the impact of STS science on students, and therefore must be carefully weighed in light of the teacher variable.

STS science instruction has relevance to a student's everyday world. Evaluators wanting to assess student learning are often dismayed by the fact that the STS content to be learned in school (formal education) can also be learned in the everyday world (informal instruction), notably from the mass media (Solomon, Chapter 10; Lucas, Chapter 11). The mass media is another "menacing" variable influencing student outcomes of some STS projects. Hence, if a student becomes more knowledgeable about STS content, was it due to good STS science teaching or to an out-of-school source? This problem is acknowledged, but a solution is illusive.

Another issue arises from an STS course's explicit connection to a student's everyday world. Although such relevance usually enhances student motivation, and therefore achievement (Mesaros, 1988), relevant contexts may to some extent obfuscate the acquisition of science content and solving science problems (Solomon, 1987). Students tend to experience difficulty when mentally moving between the theoretical world of pure science concepts, characterized by logical reasoning with evidence, and their everyday world of commonsense concepts, characterized by social interactions and consensus (Lijnse, 1990; Solomon, 1984). If STS science expects students to learn the science content in enough depth to use it in everyday situations (rather than to memorize it for the benefit of an examination), then STS science has taken on a *much more rigorous task* than traditional science has, as evidenced by the work reported by Harrie Eijkelhof in Chapter 19. The central issue comparing STS and traditional science teaching, therefore, could be *task difficulty,* rather than STS content. The problem requires more research before valid conclusions can be reached.

STS instruction is comprised of both lesson content and teaching methods. Although content and methods are often interdependent, this chapter emphasizes the content aspect of STS instruction. Before embarking on this task, however, some succinct comments on the research on teaching methods are relevant.

STS TEACHING METHODS

Traditional science teaching methods tend to be characterized by convergent thinking and lecture-demonstrations. STS science instruction, on the other hand, demands a wider repertoire of teaching strategies such as divergent thinking, small group work, student-centered class discussion, problem solving, simulations, decision making, controversies, debating, and using the media and other community resources (Aikenhead, 1988b; Solomon, 1989, 1993).

There is little research identifying the effects of these teaching strategies on STS instruction itself (Hofstein et al., 1988). Notable exceptions include the *Discussion of Issues in School Science* (DISS) project, a research program based on small-group work and applied specifically to STS science content (Solomon, 1988a). The DISS project documented students' capabilities at conducting effective small-group discussions on science-related social issues. (Its effect on student understanding of STS content is reviewed later in this chapter.) Byrne and Johnstone (1988) generalized the efficacy of small-group discussions. In an article entitled "How to Make Science Relevant," they concluded: "It is the achievement of interactivity, rather than the exact format, whether it be simulation, group discussion or role playing, which is central to attitude development" (p. 44). Attitude development is a key goal of STS instruction. Attitude was embodied in two of the four common aims of STS education stated at the outset of this chapter.

Interactive learning approaches are often identified as being essential to STS science instruction (Solomon, 1987, 1993). The research evidence suggests the following (Byrne & Johnstone, 1988):

1. In terms of learning science content, simulations and educational games can be just as effective as traditional methods. In terms of developing positive attitudes, simulations and games can be far more effective than traditional methods.
2. In terms of attitude development, the strategies of role playing, discussion and decision making can be highly effective.
3. "Group discussion can stimulate thought and interest and develop greater commitment on the part of the student." (p. 45)
4. In terms of promoting an understanding of the processes of science, an analysis and evaluation of historical case studies can be effective.

According to students engaged in simulations in the STS textbook *Logical Reasoning in Science & Technology* (described in Chapter 20), the concrete connections between the academic science content and the student's everyday world made the academic science more interesting to learn for 80% of the students, compared with 8% who found simulations of little or no value.

IMPACT OF *STS* ON STUDENTS

What benefits accrue to students studying science through STS? The impact of STS instruction can be assessed by comparing its outcomes with the objectives of an STS project, or with the outcomes of traditional science instruction.

Based on a decade of research on National Science Foundation funded curriculum projects in the United States, Welch (1969) warned of the pitfalls of "horse race" evaluation studies (course X versus course Y). Such studies tend to follow experimental designs too simple to capture the complexity of classroom life and student outcomes. Nevertheless, consumers of research on STS tend to have a dichotomous view: status quo versus the new development. This chapter takes seriously the dichotomous view of those consumers.

Before comparing the success of STS and traditional science teaching, it is enlightening to remind ourselves how successful traditional science teaching is at accomplishing its primary objective: to produce knowledgeable human resources for the professions of science and engineering and for a democratic public working in an industrialized society (Majumdar et al., 1991). Deficiencies in the outcome of traditional science teaching have been the cornerstone of arguments supporting an STS approach. Part I of this book is replete with evidence and arguments; for instance, traditional science education has failed to produce an adequate and qualified cadre of scientists and engineers in the United Kingdom and United States (Bondi, 1985; Hurd, 1989). Traditional science education has succeeded at mystifying the general public by sustaining such myths as objectivity and certain truth (Aikenhead & Ryan, 1989; Bondi, 1985; Lederman, 1992; Millar & Wynne, 1988).

What is the impact of STS science instruction? The question is answered in three parts: (1) by examining the research into the attainment of STS objectives, (2) by examining the research into formative evaluation of STS projects, and (3) by examining some research into the current status of traditional science education related to STS.

Attaining STS Objectives

Table 5.1 in Chapter 5 describes a variety of approaches to teaching science through STS. The following review is generally organized according to the categories in Table 5.1, beginning with an examination of the effects of "category 1" teaching, then moving to instances of higher category instruction, and ending with "other studies"—those difficult to categorize according to Table 5.1.

Category 1 (Motivation by *STS* Context). No research has been conducted in STS classrooms where STS content is used for the sole purpose of motivating students to learn science content. In terms of learning STS content such as the nature of science and scientists, however, research from the 1960s indicates that such content is not generally learned by students when it is not evaluated by the teacher (Aikenhead, 1973; Hurd, 1971). Similarly, in a category 1 STS class, little learning about STS content would be expected because that content is not evaluated. Motivation by STS content holds little promise for meeting STS objectives in science classrooms.

Category 2 (Casual Infusion of *STS* Content). Two studies into the effects of introducing STS vignettes (short anecdotes related to STS content along with periodic class discussions) into science classrooms showed no effect on students (Rubba, McGuyer, & Wahlund, 1991). After 6 weeks, experimental and control students were similar in their (1) awareness of current

STS issues, (2) perceived importance assigned to current STS issues, and (3) achievement on teacher-designed tests on science content.

Category 3 (Purposeful Infusion of *STS* Content). Six *History of Science Cases* (Klopfer, 1966) were developed to teach a number of concepts dealing with social issues both external and internal to science (the nature of science, scientists, and the scientific enterprise). In an exemplar research study, two history of science case studies in each of biology, chemistry, and physics were inserted into regular science classrooms, randomly chosen in the United States (Klopfer & Cooley, 1963). Each case study required 2 weeks of instruction time. Students were tested on the STS content as well as on the science content of the course. Four weeks of instruction, infused at different times during the school year, made a significant difference to students' attaining the modules' STS objectives. At the same time, the "experimental" STS group fared equally well on standard subject matter tests compared with the control group, which made no gains on STS content.

History of science infusions into British middle years classrooms showed similar results. Teachers chose between 6 and 13 units from *Exploring the Nature of Science* (Solomon, 1991) and used short activities (poster making, experiment, role playing, etc.) to encourage students to examine the text carefully (Solomon, Duveen, Scot, & McCarthy, 1992). Data from student interviews and questionnaires documented the conclusion:

> [T]here is a significant move away from serendipitous empiricism and toward an appreciation of the interactive nature of experiment and theory. Experiments are seen to be designed for trying out explanations, and hence carry an expectation about what may happen. The theories the pupils hold are now less likely to be just facts, but like experiments, are related to explanation or prediction. (p. 418)

Moreover, preliminary evidence suggested that "studying the history of a change in theory may make the process of conceptual change a little easier" (p. 419).

Harvard's *Project Physics* course (Holton, Rutherford, & Watson, 1970) was designed to systematically weave humanistic objectives into a traditional physics offering. Its major goal was to reverse the downward trend in physics enrollments in high schools (Welch, 1973). To accomplish this goal, humanistic aspects of physics (mostly social issues internal to the scientific community) were systematically infused into the course and were systematically assessed. Extensive, state-of-the-art, evaluation studies showed that *Project Physics* did reverse enrollment trends (Welch & Rothman, 1968) and did cause students to achieve a number of STS objectives not achieved by other physics curricula claiming similar objectives (Aikenhead, 1974). Achievement on standard physics tests did not suffer.

Thomas (1985) showed how grade 10 students developed the skill of structuring arguments on social issues related to science. By studying two STS modules infused into a regular science class, most students learned to find evidence, warrants, and rebuttals in arguments related to nuclear war and to global starvation. Critical thinking can be improved with STS materials that emphasize those skills.

A U.S. high school chemistry course was changed into an STS chemistry course by implementing a 6-week instructional model that involved students in chemistry-related social issues (Pedersen, 1992). Students were required to define a social issue, research it, construct an arguable position, participate in a "public meeting," attempt to reach a consensus with the class, and finally, involve themselves in a course of action. The experimental (STS) students significantly out-performed the control students in three areas: attitudes toward science, anxiety toward science, and problem-solving perceptions. In a fourth area, chemistry achievement, there was no difference between the two groups.

In a different American study, grade 7 students participated in a 4-week STS biology module that dealt with four aspects of STS instruction (Wiesenmayer & Rubba, 1990): the STS foundation level, the STS issue awareness level, the STS issues investigation level, and the STS action skills development level. Students who studied the module made significant gains in achievement on STS content compared with students who did not. The achievement was assessed with respect to: (1) knowledge about the interactions among science, technology, and society; (2) awareness of STS issues; (3) knowledge and skills necessary to investigate STS issues; and (4) knowledge about citizenship actions.

Another technique that infuses STS content purposefully into a science course is the use of video "news clips" to initiate student-centered, small-group discussions throughout a science course (Solomon, 1993). Students participating in the *Discussion of Issues in School Science* (DISS) project demonstrated the skills needed to engage in meaningful discussions without the help of an adult. As a result of those discussions, students became more cognizant of their civic responsibility to be more self-reliant in making up their own minds on an issue, rather than relying on an expert. Similar results, established through a decade of research in the United States, show that students' social responsibility is fostered by STS courses that (1) involve students in investigating STS issues, and (2) attempt to resolve those issues (Ramsay, 1993; Rubba, 1987a).

In summary, purposeful infusion of STS content can make a difference to attaining STS objectives without jeopardizing student achievement in traditional science content. On the other hand, students who follow a traditional syllabus show no advantage in science content achievement and make negligible gains on STS content achievement.

Category 4 (Singular Discipline Through *STS* Content). STS materials classified as category 4 or above depart from the structure and the focus of traditional science courses. These STS materials usually raise questions in teachers' minds about what is worth teaching in the name of science (Roberts, 1988). Science teachers, individually or by committee, are the ones who normally decide what materials to use in their classroom. These decisions are seldom based on the results of research with students. This situation does not encourage STS developers to evaluate their materials in terms of the impact materials have on students.

All too often the responsibility for assessing the impact of the STS materials resides in the hands of one person or a small team of educators. Understandably, their priority is the quality and acceptance of the materials. These developers channel all their energy into product quality and the politics of implementing nontraditional materials. Thus, little attention is normally given to research.

A notable exception is the PLON project (Eijkelhof & Kortland, 1988). Extensive research on the project's 40 modules included a systematic analysis of student's reception to those materials. In a later section of this chapter, more will be said about the PLON research that informed the revision process used to improve the modules (formative evaluation). In Chapter 19, Eijkelhof describes a specific PLON study that demonstrates exemplary research methods and yields enlightening results. In terms of the present topic—attaining STS objectives—the following summary is offered. Similar to the results of Harvard's *Project Physics,* PLON students attained the STS objectives to a reasonable degree while showing no differences, compared with non-PLON students, on standard physics questions (Eijkelhof & Lijnse, 1988). One of PLON's objectives, to promote the use of physics knowledge in everyday situations, was partly accomplished. Traditional physics courses that include similar rhetoric in their objectives offer no empirical evidence to support their claim that their materials successfully promote the application of physics to everyday situations. One of the most significant outcomes of the PLON research program is its in-depth research on learning physics subject matter and on determining what subject matter should be taught from an STS point of view (Eijkelhof, 1990; Eijkelhof & Lijnse, 1988).

The *Chemical Education for Public Understanding Program* (CEPUP), (1991) developed eight STS modules for students aged 11 to 15. As described by Herb Thier and Barbara Nagle in Chapter 8, the modules highlight chemical concepts and processes associated with current societal issues such as toxic waste, risk analysis, groundwater, and determining threshold limits. CEPUP has undertaken several comprehensive research studies to assess its impact on students. CEPUP students made significant gains, compared with their non-CEPUP counterparts, in the following areas (Kelly, 1991): perceived empowerment in dealing with chemical pollution; favorable perceptions of the chemical industry and of the media; and understanding risk, the nature of science, the processes of science, and science concepts.

Categories 5 to 8 (Science Through *STS* Content). Classroom materials classified in categories 5 to 8 tend to integrate various science disciplines. Although the issue of integration itself is not addressed in this chapter, one general statement can be made: Most researchers agree that students in integrated courses learn the traditional subject matter as well as students in discipline-centered courses, but that high school teachers are uncomfortable with integrated science (Chamberlain, 1985).

Given their cool reaction to integrated science, traditional science teachers are even more uncomfortable with the degree of STS content found in courses classified in categories 5 to 8. As a result, developers are usually preoccupied with the concerns of science teachers, which leaves little time for research with students. Some studies, however, have demonstrated that STS objectives can be achieved.

Zoller and colleagues (1990) looked at the impact of Science and Technology 11 (a category 6 course) on grade 11 students, with respect to five major objectives. The selection of these objectives for assessment was determined by the availability of pertinent and valid evaluation questions. The researchers used 4 questions from a pool of 114 STS items known as *Views on Science-Technology-Society,* VOSTS (Aikenhead & Ryan, 1992; Aikenhead, Ryan, & Fleming, 1989), and conducted a matched-pair investigation of students who had studied the new STS course (ST11) and those who had not (non-ST11). Rather than treat a student's choice as correct or incorrect, Zoller and associates (1990) analyzed the students' response patterns. The researchers identified the ideas that students chose in response to a VOSTS item. For three of the four VOSTS items, the ST11 student responses were significantly closer to the course's objectives than were the non-ST11 responses. The Science and Technology 11 course made a significant difference. In a follow-up study, Zoller, Donn, Wild, and Beckett (1991) corroborated these results, and more. By comparing students' views with their teachers' views, Zoller and colleagues (1991) concluded that the positive effects of the STS course were achieved through "education" rather than by "indoctrination."

It is interesting to see what happened with the fourth VOSTS item, the one that showed no difference between the ST11 and non-ST11 groups

(Zoller et al., 1990). Students predominantly form their views on STS topics from their culture, particularly the mass media (Aikenhead, 1988a; Solomon, 1993). STS instruction would be expected to "correct" any naive or out-of-date ideas picked up by students from the mass media. However, if an objective of an STS course harbored the same misconception found in the media, then the instruction would not be likely to alter students' views on that topic. Such was the case with Science and Technology 11 and its definition of technology: the application of the concepts and principles of science. Contrary to this view, technology is not applied science (Collingridge, 1989; Fleming, 1989; Layton, Chapter 4; Snow, 1987). If a simplification must be stated it should read, "Science is applied technology" (McGinn, 1991). The instances of technology-driven science far exceed the instances of science-driven technology (Ziman, 1984). Science and Technology 11 perpetuated an outdated concept of technology, a concept embraced by the media and shared by the teachers of the course (Zoller et al., 1991). Therefore, the STS course did not influence students' misconceptions of technology.

Another full-year course, *Science: A Way of Knowing* (Aikenhead & Fleming, 1975), claimed to teach STS objectives. Research showed that a number of these objectives were attained (those that received explicit and repeated emphasis in the course); for example: "the distinction between science and technology, the relationships between the two, the importance of open communication to the advancement of science, the tentativeness of scientific knowledge, some human characteristics of scientists, and the role of science in helping to solve social problems" (Aikenhead, 1979b, p. 127). The effect of *Science: A Way of Knowing* on student achievement on traditional science content is discussed in a later section, "Fear Over Students Missing Essential Science Content."

STS and environmental education (EE) share common objectives. For instance, STS and EE "will empower [students] to use the values, habits, knowledge, and processes of both science and democracy to cope with both personal and societal decisions" (Ramsay, 1993, p. 243). In his review of U.S. research on EE, Ramsay (1993) amplified Kortland's (1992) findings when he concluded that little impact on students' social responsibility occurred as a result of STS instruction that used sterile simulations and emphasized science knowledge (for example, *ChemCom,* Chapter 9), but significant impact occurred when students investigated real STS issues and attempted to resolve them.

An intriguing result emerged during an evaluation of *Salters' Science* (UYSEG, 1991), a series of courses for 11- to 16-year-olds. *Salters' Science* puts great emphasis on learning science in everyday contexts. Although their modicum of STS content was not assessed per se, Ramsden (1992) found that students' interest and enthusiasm were significantly peaked by the courses. But she also discovered that the more students enjoyed *Salters' Science,* the less they perceived it to be science.

The objective to improve the decision-making capability of students is often prominent for most STS instruction in categories 5 to 8. Decision making is a very complex process (Aikenhead, 1985a; Piel, 1993). The interplay between science content and decision making on a science-related social issue has been explored, but the results are conflicting (Fleming, 1986; Iozzi, 1979; Solomon, 1988b). Yet a recent in-depth analysis of small-group discussions over STS issues (part of the DISS project) revealed a subtle way in which students incorporate scientific content into their discussions (Solomon, 1993). Similar subtleties were observed by teachers who used student-centered discussion when teaching *Logical Reasoning in Science & Technology* (Chapter 20). Promising new avenues of research into student decision making were recently explored by the Dutch Environmental Education project (Kortland, 1992). This work is reviewed below in the section "Formative Evaluation."

In summary, although more research is required to illuminate the impact of recently developed STS science courses classified in categories 5 to 8 of Table 5.1 (Chapter 5), initial evidence encourages the view that many of the more modest STS content objectives can be attained, including the improvement of students' orientation toward social responsibility.

Other Studies. Written responses to a British examination for an STS syllabus were analyzed by Solomon (1988a), who concluded that explicit instruction can make a difference, but that the STS content is rather difficult and challenging to students.

Interesting results are found in an extensive research program that dealt with the impact of STS instruction on students in grades 4 to 9 taught by teachers who participated in the Iowa Chautauqua project (Yager, Blunck, Binadja, McComas, & Penick, 1988). Compared with the projects reviewed above, the STS instruction stemmed much more from a grassroots approach to developing STS instructional materials. (See Bill Williams, Chapter 6, for more details about the grassroots approach.) During 4 years of research, about 300 teachers participated in 3-day workshops on STS. Each workshop emphasized that STS instruction is oriented toward local social issues. The participating teachers designed a 1-month STS module on their own. Their own students and a control group of students were pretested and posttested on a battery of instruments and evaluation techniques (Yager & McCormack, 1989). These teachers returned for a second workshop to share their results with peers. Some teachers became leaders in subsequent workshops. STS instruction was shown to have a significantly greater impact on students in the following ways:

1. Students become more capable at applying science concepts to new situations (Varella, 1992).
2. "Students are better able to apply information, to relate information to other situations, to act independently, and to make decisions" (Yager, et al., 1988, p. 7).
3. Attitudes towards science classes, towards the perceived usefulness of those classes, and toward science careers were much more positive (Banerjee & Yager, 1992).
4. Students displayed much higher frequencies of creativity (Penick, 1992).
5. Science process skills achievement was two to three times better in STS classes (Binadja, 1992).

At the same time, the acquisition of traditional science content was the same or significantly better for STS students compared with the control group (Meyers, 1992). "STS efforts may be effective where traditional approaches fail because students start with their own problems, collect their own data, apply it to their problems, and make decisions regarding their action" (Binadja, 1992, pp. 99–100). In summary, STS instruction consistently appears to augment student achievement on STS content—without interfering significantly with the acquisition of traditional science subject matter. This conclusion raises the question: How much time can be spent on STS content before making a difference to student achievement on traditional science content? The question is answered later, in the section "Fear Over Students Missing Essential Science Content."

Formative Evaluation to STS

The research reviewed above was designed predominantly to inform decision makers and science educators about the effects of STS science instruction (summative evaluation). Another function of research is to inform the developers of classroom materials on how to modify those materials in order to better achieve the developers' objectives (formative evaluation). Formative evaluation provides systematic feedback to the designers and writers of STS materials. The formative studies reviewed here demonstrate the impact that STS instruction can have on students.

One finds very few formative evaluation studies published in the literature because information on obsolete modules is of little interest to other educators, or because few STS projects have the resources to collect evidence on student outcomes. Too often developers rely simply on teacher reaction to what went smoothly and what caused problems in the pilot classrooms. Three projects, however, PLON, *Environmental Education* (EE), and the *New York Science Technology and Society Project* (NYSTEP), have published formative evaluation data.

The PLON and Environmental Education projects are exemplars for all STS educators. They demonstrate how modifications can be made to materials by systematically collecting feedback from students and acting on that feedback. For example, early versions of PLON materials tended to be too ambitious and too general in focus and these versions were completely rewritten (Eijkelhof & Lijnse, 1988). Some of the details on the PLON formative evaluation are presented by Harrie Eijkelhof in Chapter 19.

The Environmental Education projects' formative evaluation collected data on students' ability to engage in decision making (Kortland, 1992). Three criteria were used to assess student achievement: range of alternatives, depth of argumentation, and weighting of arguments. Evidence clearly showed a significant improvement in the range of alternatives that students would generate, but no change was observed with respect to the depth and weighting criteria. Kortland also discovered a gulf between students' positive attitudes toward a decision reached and their behavioral intention to carry it out. The Dutch project represents a rare foray into assessing the STS goal of decision making.

A set of 12 STS modules produced for middle-level science courses by NYSTEP deals with science-related social issues such as waste management, epidemics, global warming, and communication. Formative evaluation data collected during field trials showed that students' awareness of societal problems increases and that the modules made an impact on the communities in which they are implemented (Cheek, 1992b).

A different approach to formative evaluation is illustrated in Chapter 20. During the development of a textbook for a grade–10 STS course, *Logical Reasoning in Science & Technology,* LoRST (Aikenhead, 1991a, 1991b), I became a participant observer in pilot classrooms, teaching the first version myself and then sitting in three classrooms watching untrained teachers teach a revised version. I revised the material again in a way suggested by students themselves, added material that students asked questions about, and reacted to problems that naturally arose in the classroom. Thus, the learning objectives of LoRST would likely be achieved to some noticeable degree as evidenced by the modifications to the textbook that were based on student achievement itself.

In summary, although the results of formative evaluation inform the revision of student materials or an accompanying teacher guide, the documentation of these results, although seldom reported, suggests that the objectives of STS materials can be attained to varying degrees. These results add weight to the findings of summative evaluation studies reported above.

Current Status Reports

STS science educators are interested in students' preconceptions related to STS content. These results serve as baseline data against which results of STS instruction can be compared when assessing education achievement. A survey of student familiarity with STS content has been used to inform policy statements about science education achievement (Bybee, Harms, Ward, & Yager, 1980; Hueftle, Rakow, & Welch, 1983). Current status studies also yield data on the effectiveness of traditional science instruction at conveying accurate STS content. Research data can pinpoint remedial content needed in STS materials. Examples are summarized here.

Canadian students' reasoned beliefs were documented on a wide range of STS topics using the STS assessment instrument *Views on Science-Technology-Society,* VOSTS (Aikenhead, Ryan, & Désautels, 1989; Aikenhead, Ryan, & Fleming, 1989; Ryan & Aikenhead, 1992). Results from these 17-year-olds indicated that: (1) views on the nature of science tend to be naive; (2) views on technology reflect the gadget orientation found in Solomon's results (1988a) in the United Kingdom, and also reflect the outdated notion prevalent in North America that technology is simply applied science; (3) the internal social aspects of science make intuitive sense to 17-year-olds; (4) the external social aspects of science are seen to be shaped to some extent by culture and the mass media; and (5) students only moderately associated social issues with science and technology. Results from VOSTS are currently being used by science educators as baseline data for their own research and teaching (for example, Visavateeranon & Finley, 1993).

Research data can suggest remedial content needed in STS materials. The VOSTS research in Canada revealed that half of the sample of grade 12 Saskatchewan students did not understand the meaning of "objective," as in "scientists are objective in their work" (Aikenhead, 1988a). As a consequence of this research, a section "Subjective versus Objective" was written into an STS textbook being developed at that time, *Logical Reasoning in Science & Technology* (Chapter 20).

Millar and Wynne (1988) monitored public views about science, views so naive or erroneous that they would be detrimental to interpreting or coping with STS issues. Similar negative effects on decision making have also been described by Aikenhead (1985a, 1985b), Layton (1986), and Layton and colleagues (1986), who discovered that the knowledge most useful to adults making decisions is the knowledge of the nature of science and how scientists work, rather than a compilation of scientific facts and concepts (traditional science content).

Conclusion

STS instruction can make an appreciable difference to a student's understanding of STS content (STS interactions, the nature of science and technology, and the social issues inside and outside the scientific enterprise), to a student's thinking skills, and to a student's attitude toward science. Although student understandings and attitudes can be resistent to change, the high degree of interest and enthusiasm that students express for STS instruction indicates that future developments toward STS instruction will receive encouragingly positive reaction from most students.

A favorable student response to STS instruction is one thing, the reaction of traditional science teachers is quite another issue. This chapter ends by addressing traditional teachers' fears that STS science will be detrimental to their students' future study of science.

FEAR OVER STUDENTS MISSING ESSENTIAL SCIENCE CONTENT

High school teachers who defend traditional science instruction warn that time spent on STS content diminishes preparation for university science courses. Fewer science concepts will be learned and general achievement will suffer. Is there any validity to these claims? The answer is a qualified "no."

The studies reviewed above in this chapter have shown that standardized measures of achievement on science content are virtually unaffected by infusing STS content into science courses, or by reorganizing the science content in ways defined by an STS issue.

Further studies have been undertaken in the United States to determine if high school science courses make any difference at all to students' achievement at university. In other words, what happens to students, enrolled in a university science class, who neglected to take the prerequisite science course at high school? Although this extreme case of ignoring a prerequisite is certainly not desirable, the situation does shed light on the issue of possible harm of studying science through an STS approach.

Science teachers are understandably concerned about their academically talented students' achieving successful careers in science and technology. (The "less talented" students normally do not take career-oriented university science courses.) What does a high school chemistry or physics course contribute to students' achievement in first-year university science? The research evidence is clear. The courses contribute essentially nothing.

Whether or not students study chemistry or physics in high school has no correlation with their science achievement in first-year university or college (Champagne & Klopfer, 1982; McCammon, Golden, & Wuensch, 1988; Tanaka & Taigen, 1986). On the other hand, students' ability in algebra and critical thinking, and their school ranking, does have a weak correlation with university science achievement. Success in university science courses relies much more on personal attributes such as study habits, curiosity, motivation, creativity, and intellectual brightness (Razali, 1986; Stuart, 1977; Susilo, 1976; Thomson, 1975). Although this result is often acknowledged by university science professors, it would seem to be a well-kept secret from high school science teachers. Most traditional science teachers will tell you that the skills and knowledge measured by biology, chemistry, and physics achievement tests are essential for success in university science courses (Ogden, 1975). They are wrong (Razali, 1986; Stuart, 1977; Susilo, 1976; Thomson, 1975).

Two experiments were undertaken in which first-year university physics and chemistry courses were taught to the more academically talented students, of whom half had not taken the prerequisite in high school (Yager & Krajcik, 1989; and Yager, Snider, & Krajcik, 1988). In both experiments, students who had not taken the prerequisite course performed as well as those who had, based on their final examination mark, their course grade, and their attitude toward the subject. The only difference was the amount of tutorial time sought out by students. Those who had not taken the high school prerequisite asked for more help. In a related study, Stanley and Stanley (1986) were able to condense the United States' full-year, high school biology, chemistry, and physics courses into 3-week summer school offerings for academically bright students.

Following the field trial of *Science: A Way of Knowing* (Aikenhead & Fleming, 1975; a category 7 STS course, Table 5.1 in Chapter 5), achievement comparisons were made between students who had studied the STS course in grade 10, and those who had taken the regular grade 10 offering (a traditional content prerequisite to 11th grade biology, chemistry, and physics at that school). Comparisons were made the year following. By November of their 11th grade science course (biology, chemistry, and/or physics), the achievement difference between students who had studied the STS course and those who had not was indistinguishable (Aikenhead, 1982). These results were particularly threatening to the traditional science teachers at that school. They successfully lobbied to have the STS course discontinued in their school.

The traditional high school science courses in North America do not stand up to scrutiny as well as one might expect. The research findings reviewed here raise a number of questions about high school science teaching, about standard measures of achievement, about first-year university science courses, and about science teachers' perceived mission in science education. The research does indicate that high school science prerequisites to university science courses in North America are not as sacred as science teachers hold them to be. Therefore, the fear that STS content will significantly diminish the amount of science content acquired by students, and thus compromise their achievement at university, seems to be unwarranted.

Obviously the issue of achievement in science is much more complex than portrayed by the succinct review here. Other factors, such as specialization over a number of years, seem to affect student achievement, as the IEA's Second International Science Study shows (Keeves, 1990).

The conclusion to be drawn here is not that high school science is irrelevant to university achievement. But rather, that high school science, as narrowly defined by teachers representing a traditional perspective, is irrelevant to the objectives espoused by its defenders. The fear over students missing essential science content may be a manifestation of a fear of losing control over the science curriculum, over students, or more personally, it may be a fear of losing one's vision of oneself as a science teacher whose mission is to socialize high school students into a scientific discipline. Such a mission offers a seductive raison d'être for maintaining the status quo in science education (Mitschke, 1993).

CONCLUSIONS

Only a small proportion of STS science projects have systematically investigated the attainment of objectives related to traditional and STS content. Over the years, however, numerous studies have been published. The accumulated body of research allows one to interpolate, extrapolate, and synthesize the following conclusions:

1. Students in STS classes (compared with traditional science classes) can significantly improve their understanding of the social issues both external and internal to science, and the interactions among science, technology, and society—depending on what content is emphasized and evaluated by the teacher.
2. Students in STS classes (compared with traditional science classes) can significantly improve their attitudes toward science, toward science classes, and toward learning, as a result of STS content and as

a result of STS teaching methods that incorporate student interactivity.

3. Students in STS classes (compared with traditional science classes) can make modest but significant gains in thinking skills such as applying science content to everyday situations, critical and creative thinking, and decision making, as long as these skills are explicitly practiced and evaluated.
4. Student achievement on traditional subject matter at the next level of science education (at a higher grade level or at university) will not be significantly compromised by teaching science through STS. This result is particularly true for academically more talented students.
5. Students can benefit from studying science through an STS approach, provided: (a) the instruction is at least a category 3 type or higher (Table 5.1 in Chapter 5); (b) appropriate classroom materials are available; and (c) a teacher's orientation toward science instruction is in reasonable synchrony with the STS approach expected to be used, as detailed by Peter Fensham and Deborah Corrigan in Chapter 18.

Yes, students can benefit from learning science through STS, and they benefit quite consistently over their counterparts in traditional science classes. STS science education, therefore, provides a demonstrative advantage to students wishing to improve their scientific literacy.

In terms of improving STS education, the potency of research and development projects was evident in the literature. Such projects hold promise for educators who wish to improve or establish STS programs.

Chapter 17

Toward a Map of Problems in STS Research

Joan Solomon

The aim of this chapter is not to list the research in this field—that is well covered in Chapter 16. Nor will I give full accounts of the research that I have carried out over the last 10 years and reported elsewhere. Instead, I will use more general theories to tease out some of the complexities that may be encountered in this kind of research.

TEACHING AND CURRICULUM DEVELOPMENT AS RESEARCH

When a school subject is very new its reality comes into being through the act of teaching. That takes the craft of the teacher to its very pinnacle of creativity, and the STS movement owes more than it can repay to those who first struggled to get across to their students not just the run-of-the-mill science concepts, but a course of lessons that would reflect their image of science placed firmly in its social context. Delivering such innovative lessons requires an openness to feedback from the students that almost amounts to a research paradigm in itself. I write "almost" because research in science education, as in science itself, is totally dependent upon the communal element. It needs to speak to teachers and educators, and place itself within a written tradition of educational innovation.

In Chapter 20, Glen Aikenhead provides a splendid example of just such primary research in which one can hear the students' voices. The paper I wrote some 10 years ago (Solomon, 1985) about how my students tried to understand the social context of energy, is a contribution to this genre. Others may be found, but they are few in number for reasons that include the sad and continuing divide between those who teach and those who write about teaching. The closely observed classroom research I have in mind can do more than provide feedback for the teacher, or informal validation of the new course. It could illustrate how it feels to be a student learning this new subject.

EVALUATION OF *STS* AS SCIENCE EDUCATION

Research that attempts to validate STS as a complete science program is most frequently found in the United States and Canada where there is an either-or policy of curriculum design. An STS program, if adopted, will be the only vehicle for learning science. It will shoulder the sole responsibility of delivering scientific understanding, as well as fulfilling the more specific goals of STS, such as understanding the nature of science and technology and their concepts; their interactions with society and decision making about them in a social context.

The first of these provides a two-horse race with conventional science teaching and lends itself to analysis by the standard testing of student outcomes. However nothing is ever simple in so complex an area as education. Bruner (1966) reminds us that

> evaluation should examine not only the product or content of learning, but also the process by which the child gets, or fails to get to mastery of materials. . . .Content cannot be divorced from pedagogy. For it is the pedagogy which leads the child to treat content in critical ways that develop and express his (sic) skills and values. (p. 164)

This suggests that content, the learning process, and the acquisition of relevant skills should all play a part in evaluation. This is a harsh requirement but essential of all research (see Merton, 1942, for the norm of "organized skepticism"). Credence in our work, for which we struggle, is not withheld because the data are in doubt or because any chicanery is imputed, but because of a lack of real understanding of how the claimed learning success came about. Only if we can show in detail how cognitive hurdles were overcome and how the tools that our students used to solve questions were taught and learned will our evaluation convince an audience of peers.

The point I am making has been made before, and better, by the British social historian W. G. Runciman, who wrote in 1989 that social science research consists of four essential and interwoven strands: reportage, description, explanation, and evaluation (in the sense of exposing values). In our case, the first of these may comprise pre- and posttests, or the ticking of boxes, or the grading of examinations, but none of these methods will be enough in itself. They would omit the necessary ethnographic description of the students' experiences, as well as our own grounded explanation or theory of learning by which to analyze and illuminate the steps of the learning process. This provides another suggestion for a program of STS research.

RESEARCH INTO THE SOCIAL RELATIONS OF *STS*

STS programs also have other goals. They claim to teach about the social relationship of science, and in this they sometimes meet incomprehension and skepticism. The stratospheric language of the aims of some STS courses do not help in directing research, but the social dimensions can be put in simpler terms. Understanding the "interactions of science and technology with society" means that students will need to explore how innovations and environmental changes effect different sectors of their own society and also, perhaps, another society in a contrasting part of the world. The teaching strategies employed may include role play and simulations, but the essential ingredient is an attempt to see the issue from the various points of view of those who may be differently affected by it.

In 1988, I conducted a small inquiry into this aspect of students' learning (Solomon, 1988a) and concluded that the skill of being able to see an issue (in vitro fertilization) from different perspectives was not too demanding for the common run of students. Since that time, I have had several opportunities to observe the surprising natural skill of students in this respect. Indeed, the task that we science teachers routinely present to our students of learning to manipulate a mental model (for example, a gas as an assemblage of colliding molecules, or yeast as a collection of living and respiring microorganisms) is found to be far more taxing. Manipulating a social model is like anticipating how a friend or parent might react to the news you are about to break to them, and is a life skill learned by quite young children. In the DISS project (Solomon, 1992c), and in some more recent work on how British pupils come to understand the nature of science, I have been impressed by how often students chose to use the vehicle of social role-play in order to work out an understanding of some more abstract point. In the following example, a girl of average ability (aged 15) interviewed about the role of theory in science, has been asked if she thinks that scientists use their imagination.

> If they, if they come up with some experiment that they have proven and they want to show it to the other people, and they've got to, like, to show a diagram or something that they think might be true, and they want to try to bring it to other people, and they draw a diagram and stuff and say "This is what I imagine. . . ."

In terms of Cheek's (1992c) recent analysis of the disciplines that build up STS, this extract provides the clearest possible example of constructivism at work, even though it is a social process that is being modeled, rather than a scientific one. There are two possible approaches to constructivism within education: the "conceptual change" program, and the "two domain" problem (Schutz & Luckmann, 1973; Solomon, 1984). In the first case, students are thought to throw over their old theory completely in favor of a new one as soon as they have been convinced by experimental evidence and teacher exposition that the new theory has proved its fruitfulness and worth (Strike & Posner, 1982). In the two domain framework the students are not expected to relinquish the phrases and meanings of the more familiar everyday set of ideas. Instead they accept an additional kind of knowledge in their science lessons and learn to move between the two systems of knowing—the life-world and the scientific—each with its own way of thinking as well as content. In STS a flexibility of epistemic position is quite central to the learning process. Our students have to learn their scientific knowledge in the context of people's everyday lives. Without flexible movements between domains our attempts to help our students think about justice from different perspectives (Rawls, 1971) will be doomed to fail.

WHAT ATTITUDE CHANGES SHOULD WE SEEK?

There is nothing at all novel in searching for attitude changes as outcome in STS research. It has permeated the literature on small-scale research in environmental education (Iozzi, 1984) and is now rapidly taking over in the STS field (Rubba & Wiesenmayer, 1985b). The reason for this is easy to see. The issues on which many STS courses are based—nuclear power, chemical wastes, and the use of animals in experiments, to name but a few—are exposed in the media and argued with some passion. Teachers would naturally like to know if the science teaching that they bring to bear on these emotive issues—about nuclear radiation or genetic disease—helps students to change their attitudes toward the related issues. However, there are some serious general problems with this kind of research that can be listed under three main headings.

First, we need to understand what is meant by an attitude. In traditional psychology it is "a learned disposition to respond consistently with respect to a given object" (Fishbein, 1975, p. 59). This begs a lot of questions. Are attitudes learned? Do they always predispose to action or response? In an attempt to clarify this the authors claim that beliefs, which are themselves links between objects and attributes, affect attitudes that then influence intentions that produce behavioral outcomes. It seems a very rational linear scheme, even when the influence of subjective norms on intentions are taken into account, and one that might be easily adapted to evaluate STS courses. But does it really describe the roots of human behavior?

Both Vygotsky (1978) and Moscovici (1976) have emphasized, in their own distinctive ways, the social nature of attitude formation and consolidation. For them attitudes are constructed in the context of those who are interacting together; questions and ideas are continually being tested and modified in line with impressions formed about the others present—those who impress with new knowledge and those who stick stubbornly to familiar ideas. Understanding what attitude means in these terms is hard to square with results calculated from a set of questionnaire responses. Although it may be true that filling in boxes is a social interaction between the setter and the respondent, it is clearly a weak one.

The second problem concerns the uncertain route from knowledge to attitude. The STS teacher may offer information about the polluting effects of vehicular emissions, or demonstrate the shielding of nuclear radiation by concrete, but this knowledge is often not enough to ensure a change of attitude. It is well known, for example, that those who have a morbid fear of spiders can be assured that they can do no harm to them, and yet continue to be afraid. Our attitudes are only partially related to the information or misinformation that we hold. In the DISS project (Solomon, 1992c) some science-based social issues seemed to be more subject to attitude change than were others (for example, more in the use of animals in experiments, less on questions of pollution). Both groups of students, one that was studying STS and taking part in small group discussions, and a control group of the same age that did not, showed this variation. However, on topics that had been discussed the experimental group always displayed more change of attitude than did the control group. Thus it may be a mistake to judge the effectiveness of an STS course by a simple measurement of students' attitudes, without reference to a control.

The third question is embarrassingly direct. Which attitude changes do we want to promote, and why? It is part of the high ideology of teaching that we aim to develop in our students the power of their own thinking—their capacity to inquire and to resolve. In Britain the movement to stimulate independent thought through classroom discussion is linked with the name of Lawrence Stenhouse; in the United States Jerome Bruner holds a similar position. Both might find the attempt to evaluate students' achievement in terms of changed attitudes to be contrary to the principles of good education. It is also very curious that the attitude outcome expected by different researchers, or at least welcomed by them after the event, shows great variation. The literature on this kind of evaluative research in terms of attitude change includes those who want the students to change to more positive (extreme) attitudes, those who want to swing them from one position to its opposite, and those who want to moderate the students' extreme views to what they hope will be a more reflective middle position. Thus almost any movement in overall attitude score may be taken as a vindication of an STS course. Is this a measurement of knowledge, or persuasion, or even of indoctrination?

BEHAVIORAL OUTCOMES AS VALIDATION

The highest aims of STS education relate to how our students will behave as citizens. Most governments hold a sadly instrumental attitude toward education and see investment in schooling not so much as a cognitive and cultural good, but as an investment in the country's future. Traditionalists translate this in terms of the nation's supply of qualified engineers and scientists who bring it economic wealth. Advocates of STS take a more pupil-centered view, and many from the Third World countries hope that a relevant science syllabus will directly improve their students' quality of life and health care. They would likely agree with sentiments like the following, which manage to combine high aims with new patterns of copying behavior: "The real test is whether these pupils will be more responsible citizens and better capable of mastering their environment" (Nganunu, 1988, p. 448). This seems to provide directly testable objectives.

Hines, Hungerford, and Tomera (1987) approached the question from the opposite direction. They studied the factors that seemed to be most influential in motivating "responsible environmental action" (measured by self-reported action). In a valuable meta-analysis of all the research carried out to date they found correlations with the students' knowledge about issues, at 0.299. Clearly this figure does not of itself establish any causal relation between having been taught about STS issues and self-reported action. Further analysis showed a higher correlation, at 0.365, between the students' belief in their capacity to bring about change and their engagement in environmental action. The authors make two important general points: that action skills do not evolve naturally from knowledge, and that in addition to knowledge and skills an individual needs to possess a desire to act. This vague but all-important factor, together with situational factors (economic, regional, and social pressures), may easily overwhelm the effects of STS teaching, if we are to take this prime example of statistical research at face value.

The uncomfortable fact for would-be researchers is that students' knowledge and attitudes, however they are assessed, do not provide a sure guide to future behavior. The DISS project obtained attitude responses, and also scores of behavior during recorded small-group discussion, but these two sets of data showed very little association (see Chapter 10). In social psychology this is no new phenomenon.

This finding is uncomfortable and challenging because it is all too easy to accept that claimed attitudes that are belied by actions, or at least not supported by them, can be of little moral worth. However, the social component of attitude and action may so change the meaning of what is done in different situations that the same attitudes simply do not apply. The "social honor" conferred by a student's act may be high in school but low in the home (Harré, 1983). This is beautifully illustrated by the words, reported by Kortland (1992), of a student who had studied a unit on the environment and agreed with the interviewer that milk bottles are more environmentally sound than are cartons. On being asked if he would then choose to take a bottle of milk home he replies: "No, because if I came home with a bottle my mother would look at me in a funny way. I'd never bring home a bottle" (p. 37).

CONCLUSION

The challenge, but also the profound interest, of this open field of research has been evident in each section of this critique. We keep finding that what has appeared to be rational and assessable is fraught with contradictions. It presents us with what is probably a prime example of the modern-postmodern debate complete with the "flight from rationalism" and "communicative hermeneutics" of which Rorty, Gadamer, and Habermas have written (see Bernstein, 1991).

In the liberal mood of the 1970s, when the STS movement in schools began to take shape, it was a premise that students could be empowered to make their own decisions if informed by the appropriate scientific knowledge. No one spoke then of moral principles or right actions, although the ideas hovered in the background. Students were to understand that different interest groups in society would see actions and effects as good or bad from different perspectives, but decisions about solution strategies could still be made. This disordered moral situation is precisely what MacIntyre (1985) addresses in his seminal work *After Virtue*. He believes that modernist moral philosophers have fundamentally misconceived the basic problem by putting moral principles and rules for action in the primary position: ". . .grounds for the authority of laws and virtues can only be discovered by entering into those relationships which constitute communities" (p. 258). To find that the high-flying objectives of STS are, after all, consonant with postmodern thinking may provide us with some satisfaction, but it offers little help in the conduct of our difficult research agenda.

Chapter 18

The Implementation of an STS Chemistry Course in Australia: A Research Perspective

Peter Fensham and Deborah Corrigan

Curriculum reform always takes place in the social milieu of schools and society, and is a response to a complex set of demands imposed on teachers and students by schools and society (Fensham, 1992). The last 30 years in Australia have seen different social demands produce a variety of curriculum changes. This chapter describes some research into the latest of these reforms in the state of Victoria. (The states and territories of Australia individually control their schools' curriculum. There is no national curriculum.) The latest Victorian reform produced several STS courses for grades 11 and 12 in chemistry, physics, biology, and geology. Our research investigated (1) chemistry teachers' conceptualizations of the mandatory STS chemistry course, (2) the extent to which these conceptualizations were shared with the writers of the new course, and (3) how the teachers' conceptualizations were modified by teaching the course for a year.

HISTORICAL BACKGROUND

During the past 30 years, four distinctly different social circumstances influenced the reforms of what originally had been an elitist, university preparatory, chemistry course offered in grades 11 and 12 in Victoria. These circumstances and reforms are summarized here.

Social demands of the 1960s (principally the call to purge out-of-date content from the chemistry curriculum) caused schools to incorporate into the chemistry syllabus highly conceptual and theoretical material traditionally found in university chemistry courses.

Social demands of the 1970s (principally, an unprecedented population influx into Australia, and the abolition of most external examinations) caused teachers to assume responsibility for determining the chemistry curriculum that best met the needs of a diverse group of students in a particular school. A school-based movement for curriculum development emerged. The chemistry textbooks for grades 11 and 12, however, were only modified slightly by inserting superficial vignettes of applied chemistry into abstract and theoretical chemistry.

A third social demand on teachers came in 1979 when the Victorian educational authorities decreed that all Higher School Certificate (HSC) grade 12 subjects would have an externally examined core and internally assessed options. The decree actually supported the school-based movement and left a legacy that legitimized teacher-initiated content, pedagogies, and assessment procedures.

A fourth reform responded to the social demands of the 1980s (principally social justice for a multicultural population, a faltering economy, and retention policies for early school leavers). It breathed new life into the school-based movement for curriculum development. Malcolm, Cole, Hogendoorn, O'Keefe, and Reid (1987) produced a government document *Science Framework* for the study of science from kindergarten through grade 10. The document very specifically promoted "science, technology, society, and personal development" as the emphasis for science education. The prevailing democratic climate supported this science-for-all position, but a stormy economic climate fostered an elite science education for the few who would be likely to succeed beyond schooling. No resolution between the two was reached.

Coincidentally, however,. the Victorian authorities established a new educational authority (the Victorian Curriculum and Assessment Board, VCAB) to develop and implement a new science program for grades 11 and 12 (the Victoria Certificate of Education, VCE). The actual work would be supervised by the Science Field of Study Committee (Science FOSC) chaired by Cliff Malcolm, principal author of the reform-minded *Science Framework* (Malcolm et al., 1987). The Science FOSC replaced the discipline-centered curriculum committees of the past. Members on the Science FOSC came from schools, universities, and the management and trade union sectors of industry. Supporters of the old elitist program comprised less than one-third of the committee. Science FOSC (1989) adopted the following STS framework:

> Science serves economic and environmental purposes as well as cultural and personal ones. Scientific research and development often extend across disciplines, in their social impact and in the way the research is conducted. The science studies recognize the interdisciplinary aspects of science, as part of the problem solving, applications and the social contexts of science. . . . The studies emphasize:

- theoretical knowledge and the processes scientists use to develop and test that knowledge;
- technology and the solution of practical problems in technological development and environmental management;
- the relationships between science and society, the ways science proceeds, the values that guide scientific activity, and the influences of scientific discoveries on values and beliefs in society. (p. 4)

This framework led to VCE Chemistry. Its implementation by chemistry teachers was investigated by the research project reported in this chapter. This research is clarified by the following short description of the new chemistry course.

VCE CHEMISTRY

VCE Chemistry was the fourth and by far the most complex curricular change that Victorian chemistry teachers have faced in the last 30 years. These teachers, however, were not unaccustomed to changes in curriculum, as evidenced by the historical background described above.

The magnitude of this latest change can be highlighted by locating VCE Chemistry in relation to various ways STS science has been categorized. In terms of John Ziman's approaches to STS (Chapter 2), VCE Chemistry exemplifies the approaches through relevance and the "problematic." Glen Aikenhead (Chapter 5, Table 5.1) classified VCE science as a category four STS course—singular discipline through STS content. One of us (Fensham, 1988) devised a three-point classification scheme for STS courses: science-controlled, technology-controlled, and society-controlled. The scheme describes the jurisdiction that decides which knowledge is worth learning in an STS curriculum. Of the four reforms to chemistry over the past 30 years, the first three were all science-controlled. VCE Chemistry should be classified as largely technology-controlled, according to the intentions of the Science FOSC. Thus, VCE Chemistry represents a significant change for Victorian teachers, most of whom have had their own chemistry education entirely in science-controlled studies.

Given the magnitude of the curriculum change for VCE Chemistry, its implementation presented a challenging phenomenon to study. In this chapter, we concentrate on the meanings associated with curricular information, and some of the processes and conditions that nurtured developers and teachers to share similar meanings of that information. It is our contention that much of the discrepancy between the intended curriculum and the taught curriculum is due to information being exchanged without its meaning being shared.

The VCE science courses have a number of structural features that impose changes on teachers:

1. The traditional year-long format was replaced by semestered courses (a student could now study single or double units in chemistry in grades 11 and 12).
2. Norm-referenced testing and teacher determined pedagogies were replaced by "work requirements" involving criterion-referenced assessment and required pedagogical experiences (for example, concept mapping and field trips).
3. The narrow traditional assessment of memory work was replaced by "common assessment tasks" that addressed the full, broad range of learning objectives (authentic assessment).
4. A specified course accompanied by one textbook was replaced by study guides and supported by a variety of text-type materials.

None of these features was familiar to chemistry teachers in Victoria.

Moreover, the target population of students had changed. VCE Chemistry would now be offered to more than 80% of each age cohort, rather than to a small proportion of academically oriented students.

A small, geographically dispersed team of writers was employed to write the *Study Design for Chemistry.* As they did so, they followed suggestions from the Science FOSC, the chemistry subcommittee, and the VCAB itself. It was curriculum design and development by committees and by dispersed writers, rather than by a curriculum project team. This organization inevitably created problems of shared meaning between the writers and the teachers in the schools, even before the teachers were remotely involved.

RESEARCH DESIGN

A number of studies have investigated the diffusion of many centrally planned curricula developed in the 1960s and 1970s in the United Kingdom and United States (Harding, Kelly, & Nicodemus, 1976). In almost all of the curriculum projects reviewed by Harding and colleagues, attention was focused on how and to what extent these curricula were accepted, adopted, or adapted. This research focus on decision making by schools or teachers is at odds with curricular innovations in the many other countries where schools and teachers have no such decision-making power because the implementation of the curriculum is mandatory (certainly the situation of VCE Chemistry). Nevertheless, the research designs employed by these studies can inform investigations into the implementation of mandatory curricula.

Three types of research methodologies were evident in the review by Harding and associates (1976). First, large scale surveys tried to ascertain which schools and what sorts of teachers chose to adopt the curriculum, and which did not. Dissemination—the flow of information and meaning about the innovation—was not treated as problematic. It was assumed that if information is received, its meaning is shared.

A second type of methodology used ethnographies in a descriptive-observational research tradition. The reports portrayed the development process, the issues the curriculum developers discussed, the key events, and the impact the innovation had on teachers as persons. These studies presented rich accounts, in the language of the participating teachers, of the questions that arose as teachers made decisions about the curriculum innovation and reflected on their attempts to put it into practice. The focus was on the variety of responses teachers made to information that they had received, but not primarily on the *meaning* that teachers attached to that information.

The third methodology also involved case studies and was concerned with an organizational analysis of a project from its inception to the year of its first regular use in schools. These studies were conducted from a more formal distance than were those in the second category. It is interesting to note the problem these researchers encountered. The very cooperation of their informants caused those informants to give the researcher information the informants thought the researcher wanted, rather than information describing what actually happened (Shipman, 1972). Nevertheless, Shipman found that qualitative analysis of his data did illustrate the different and changing perceptions held by persons involved in a curriculum innovation. This type of methodology suited our focus on shared meaning between VCE Chemistry writers and teachers.

Because the adoption of VCE Chemistry by teachers was a predetermined issue, we wished to focus on the dissemination process itself and how the meaning of its message about curricular changes was shared. Our theoretical stance included the following tenets:

1. Meanings are constructed by individual chemistry teachers using their existing cognition and affects (feelings).
2. Existing cognitions and affects are shaped by the social interactions these chemistry teachers have in common.
3. Because the construction of meaning is constrained by language and by common social interactions, the range of meanings expressed by Victoria's chemistry teachers is likely to be limited.

To understand the meaning that others attached to words or messages, it was necessary to listen not only to the responses teachers gave to our questions, but also to teachers' explanations of those responses. Clinical questioning (open-ended interviews with deliberate probing) was chosen, therefore, as the means of collecting data.

We were more interested in a *variety* of meanings that might exist, and how they might change, than we were in quantifying how many teachers held which meanings. Consequently, quota sampling (within our resources) rather than representative sampling guided the selection of teachers to be interviewed.

Our project was designed in three distinct phases. In the first phase we interviewed the writers of the *Study Design for Chemistry* to gain an appreciation of the meaning they attached to eight rather novel aspects of the new curriculum: content, contexts, focal questions, work requirements, practical work, assessment, changes in teaching, STS approach, and student learning. The writers were asked to explain new terms and to talk about changes from former practice (HSC Chemistry).

In the second phase of our research, 32 teachers from rural and metropolitan schools were interviewed about the same eight aspects. The interviews took place in 1989, 15 months before the teachers were to teach VCE Chemistry, but after they had received information about VCE Chemistry from a number of dissemination events.

The third phase involved interviews with 30 teachers (some who were part of phase 2) in late 1991 and 1992 after they had a year's experience teaching grade 11 VCE Chemistry.

FINDINGS

Three of the new aspects of VCE Chemistry are used to illustrate what was achieved by this research: context, focal questions, and work requirements.

Contexts

One way to resolve the problem of what knowledge is worth learning in an STS science course is to talk about concepts in context (Eijkelhof & Kortland, 1987)—the social or technological contexts associated with the science content, and in which the science content could be embedded; for example, diet, textiles, health, pollution. The Science FCSC adopted this notion of contexts, but its members did not agree on what knowledge is worth learning. Two positions emerged: (1) the knowledge content of the new course consists of concepts and the contexts in which those concepts have meaning, versus (2) the knowledge worth learning is primarily the concepts, whereas the contexts are simply ways of teaching them. This ambiguity also persisted among the writers of the new course. Their exploration of new chemistry curricula from overseas led them to find little change in the concepts being taught despite the material's strong emphasis on real life contexts. One of the writers described context as follows:

> There is little change in content. Most of the content that is there is in the present course [HSC]. The change is the context in which that content is to be taught. There is some notion of context being important in the present course [HSC]. The biosphere is an example of that. However, the emphasis placed on context is far greater now. In the old course "context" could be tagged on at the end. . . . What we want to do now is turn segments around and introduce corrosion [by] talking about Coke cans made of steel and aluminum and what's the difference and why. "Now I'll explain to you the chemistry of why that happens." . . . So a lot of emphasis here was on finding examples that were relevant, worldly, that people had experience of and might be involved at some time in making judgments about, "What sort of fabrics do you wear?" and so on. (Writer A)

The role of contexts was quite well recognized by teachers interviewed in phase 2 of our research:

> We're going to have to teach things from the opposite way because . . . you're looking from more general terms and focusing. For me it's turned inside out. It's probably about time it changed anyway. (Teacher A)

> It's great that finally now, it's not putting the chemist's viewpoint onto things. It's actually putting the kid's perspective onto things. (Teacher B)

Teachers in Victoria were acquainted with teaching a conceptual sequence to which applications were added throughout to illustrate the concepts. The common experience provided by this type of teaching seems to have facilitated the sharing of a new meaning for contexts.

Some teachers expressed less support for the contexts. They described the new course as containing the same conceptual content as the HSC but, "They have messed around with the order." Such a view is consistent with the fact that the third curriculum change in 1979 had not altered the sequence in which the concepts had been taught.

In phase 3 of the research project, 12 months after the introduction of VCE Chemistry, the teachers' responses about context varied in a manner that will become more differentiated in our discussions of "focal questions" and "work requirements" below. It should be noted that the term "context" itself was not present in the *Chemistry Study Design* document. Two examples of teachers' views of context are:

> Well you're looking at a theme approach, looking at the world situation where it is actually used. I think it is mainly looking at industry and where chemistry is used in industry. . . . I sometimes feel that it would be easy to stay teaching the old way, so to speak, without putting those linkages in. (Teacher C)

> And the context . . . certainly makes things more relevant for them, because it's the particular setting they're doing the work in. And they see the connections much better. . . . In the past, here's a little box bonding and they don't transfer it. [Now] they seem to be able to transfer it, but I don't know whether it's just this group of kids or because of the new course. (Teacher D)

Focal Questions

The intention of focal questions was to define the areas of study in each unit in terms of the Science FOSC's overarching framework of STS; for example, "What compromises must be made in the production of chemicals?" Focal questions were not intended to define chemical phenomena, knowledge, concepts, or activities, which would, however, need to be studied in order to resolve the focal questions. In this sense, then, the focal questions served as reminders to teachers of the chemistry-for-all intention of VCE Chemistry, and the link between concepts and their context, a context in which the concepts could be embedded.

In phase 1 of our study, we discovered that the writers did not distinguish between focal questions and contexts. Despite the fact that the term "contexts" is excluded from the *Chemistry Study Design* document, the writers used the term "contexts" when responding to our questions about the changes in the curriculum. Perhaps they were so immersed in the task of describing concepts and contexts that there was no need for the writers to be reminded of the concept-context link—the intended function of the focal questions. It is not surprising, therefore, that we found among teachers with whom the writers had communicated a confusion of meanings about the role of focal questions.

In phase 2 of the research project, the range of responses from the teachers varied greatly. A number described the focal questions in negative terms: "They are content headings with no elements of methodology in them." "Guidelines that are much less defined than the present course is." "A lot of garbage!" "A nightmare!" "Students won't be able to handle them, so I won't give them to them." "Students will need to be jollied along quite a bit, as it's a long stepping-stone from the content to the focal question." These comments came from teachers who projected an image of being teacher-centered or content-centered (transmitters of knowledge).

Another group of teachers, fewer in our sample, described focal questions positively and with a clear sense of their intended purposes. For example: "They are a framework to work from and provide a way into the unit. They make chemistry real and relevant. They are the kind of questions kids ask." "They develop real-life emphasis to the content and guide the methodology or ways to approach the course." Based on interview data, these teachers would be characterized as student-centered. They were more concerned with facilitating student learning than with covering the curriculum.

No prior experience of the first group of teachers had prepared them for focal questions. The second group, however, did have a conceptual structure (from their student-centered orientation) that enabled them to construct an understanding of focal questions and react to them in a positive way.

In phase 3 these divergent views on focal questions persisted, although a number of teachers began to express a much clearer purpose for focal questions. The 1-year experience also made a difference for teachers who had an initial positive view of focal questions. These teachers added considerable detail about how to use focal questions:

> Focal questions are what the student should be able to answer with varying levels of complexity. I think that is what will determine if you're assessing a student's work, not only as "Are they satisfactory?" but also the depth of understanding they have. . . . So I suppose the contexts are the way you go about [approaching a topic] and the focal questions are what you end up knowing about it. (Teacher D)

Work Requirements

The writers understood the VCE prescriptive element called "work requirements" to be an opportunity to force teachers to focus on how students learn.

> Work requirements are a little like continuous assessment in that they are intended to put requirements through the year—to force people . . . to do different types of tasks and spread them. We see it as a learning strategy—it derives from what we know of metacognition. (Writer B)

The writers used work requirements (for example, concept mapping and field trips) to tie together the structure and the philosophy of the course. A work requirement was a primary means of imposing on teachers the methodology they were to adopt. Teachers expressed resentment at first:

> It's one of the things with the new course, the work requirements in a sense prescribe methodology. That's something teachers are going to find difficult to come to terms with. Traditionally teachers don't like being told what to do. (Teacher E)

But there were some teachers who understood perfectly.

> It will be difficult for people to teach the new course out of context because there would be a significant number of work requirements that they would not be able to do because [work requirements] are geared to context. (Teacher F)

Once again a polarization arose between teachers who saw themselves as transmitters of knowledge and those who saw themselves as facilitators of learning. But after an initial year of teaching, the experience with work requirements ameliorated a great deal of the aggravation felt by some of the teachers, although a few continued to harbor a sense of outrage that professional teachers were being told how to teach. Generally, however, there was an acceptance of work requirements as useful mechanisms either for teaching or for learning (depending on one's self-image of a teacher). Two quotations illustrate each viewpoint:

> Work requirements make sure that the kids do the work. But here it's more structured, whereas in the old course, the kids did work as well, it's just that we didn't have a system whereby we actually checked how well they progressed. (Teacher C)
>
> These [work requirements] are very good. They constitute class activities such as concept maps. Other examples are the "waste" assignments where students' time has been well used. The time involved for the teacher is about the same. Work requirements foster a variety of [learning] techniques. (Teacher G)

CONCLUSIONS

The shared meaning of curriculum intentions between writers and teachers was highlighted by this study's longitudinal design. In phase 2, intentions were only words to teachers. In phase 3, teachers had 1 year of experience with the new program. This experience allowed teachers to construct meaning of the words and messages introduced in phase 2. For many teachers, as indicated by the data above, this experience brought about shifts of meaning and established more connections among those meanings.

The inclusion of rural and metropolitan samples enabled another source of meaning-making to be studied. By the end of our study, more teachers in the rural sample had changed their views, compared to their counterparts in the metropolitan center. The rural clusters were geographically determined. This brought unfamiliar teachers together to exchange views and practices. The rural teachers commented to us that their colleagues had contributed to the observed changes in their views and practices. On the other hand, in the metropolitan sample, teachers tended to check out their ideas and uncertainties with easily accessible friends among the many teachers in the Melbourne area. Hence, social interaction occurred among teachers who were likely to have similar views that reinforced meanings rather than challenged them.

In light of these findings, it is interesting to note that in 1988 a group of constructivist researchers in science education at Monash University sent an open letter of concern to the Science FOSC. In it they commended the STS framework that the committee had adopted, but warned against the development of STS science courses at the expense of a more effective and deeper learning in science education. These constructivist learning dimensions would appear to have gained in VCE Chemistry more than the intended STS content. STS content (via the focal questions) is quite unfamiliar to most science teachers with their narrow academic backgrounds. Thus, it seems harder to establish this STS content as something worth learning than it is to establish new strategies aimed at improving traditional instruction. Although the *intentions* of Science FOSC had been technology-controlled, our data above suggest that in the early implementation stage at least, the VCE Chemistry program continues to be science-controlled.

Perhaps only when external examinations in VCE Chemistry take the STS dimensions seriously (an unlikely event in the competitive university selection climate now prevailing in Victoria) will most chemistry teachers believe that the sociochemical concept of waste is as worth learning as the Lowry definition of an acid.

Chapter 19

Toward a Research Base for Teaching Ionizing Radiation in a Risk Perspective

Harrie Eijkelhof

The topic of ionizing radiation is fascinating to many people, students included. One dominant reason for this fascination seems to be the association of radiation with risks and benefits. Weart (1988) has written comprehensively about people's widespread nuclear fear. People often associate radioactivity with accidents in nuclear power stations, with nuclear bombs, with radioactive waste, and with radiation-induced cancer. On the other hand, most people know that radiation is used to detect or cure diseases, the same diseases that could be caused by radiation.

In science education, the topic of ionizing radiation is often dealt with in a purely academic manner, starting with the structure of the nucleus, continuing with the decay of unstable nuclei, properties of different types of radiation and fission reactions, and ending with a limited number of applications, such as carbon dating and the nuclear reactor system. Such an approach does not take into account the interests and feelings of students. It seems to be mainly directed toward preparing students for further academic studies.

Students' interests and feelings about ionizing radiation can be incorporated into learning materials by including questions about the acceptability of risks. Our experiences developing and testing such materials is the focus of this chapter. We discovered that an intuitive approach to writing such materials did not produce satisfactory results. Instead, success was attained by a research-based approach in which a number of diverse studies informed the writing of these new curriculum materials. This research-based approach represents a viable method for developing any STS materials.

FIRST EXPERIENCES WITH NEW CURRICULUM MATERIALS

As part of the large *Dutch Physics Curriculum Development Project,* PLON (Eijkelhof & Kortland, 1988; Lijnse, Eijkelhof, Klaassen, & Scholte, 1990), a unit was written for the senior high school with a central theme "the acceptability of risks of ionizing radiation" (PLON, 1984). The main aim was for students to learn to use scientific knowledge in reasoning about radiation risks. In terms of Glen Aikenhead's classification scheme of STS science (Chapter 5), one may place the unit in category 4. (The scientific disciplines are taught separately in this country.)

In this PLON unit, "Ionizing Radiation," students find answers to a number of basic questions developed from the unit's central theme. Questions include: What is the origin of ionizing radiation? What are its properties? How can this radiation be detected? What are its effects on human beings? What use can be made of this radiation? How can one protect oneself against radiation? How can the radiation risks be evaluated? Answers to the first three questions may seem like the physics in traditional units on this topic, but the answers are given from a different perspective: the risks that a person may face in real life. The last four questions have resulted in content that is not common in physics education: the concepts of dose and dose equivalent, subtopics such as biological effects of radiation and radiation protection, medical applications, radioactive waste, nuclear arms, and procedures for evaluating risks.

Evaluation of an early draft of this unit (Eijkelhof, 1986) showed the following results:

1. Students judged the unit to be useful, instructive, interesting, and not too difficult.
2. The unit did not cause depressive feelings among the students.
3. Girls found the unit more instructive (especially the medical aspects) but more difficult than did boys.
4. Not much improvement was measured in students' reasoning about radiation risks. Their reasoning was still dominated by common sense and lay ideas about radiation.
5. Teachers were, in general, very satisfied with the unit.

Results (1) and (5) confirmed our expectations, while result (2) refuted the assertion of some Dutch educationalists that the unit would create in students a depressive mood. Result (3) is in line with findings from research on gender issues in science education (Jörg & Wubbels, 1987). Result (4) was disappointing because the main aim of the unit was to improve student reasoning on radiation risks. At the time of my original study, hardly any research on this topic had been reported, with the exception of a German article by Riesch and Westphal (1975). Therefore, I decided to start a number of studies that would improve the unit in terms of students' reasoning about radiation risks.

OUTLINE OF THE RESEARCH

The aim of these studies was to identify the conditions that need to be fulfilled in order to promote thoughtful risk analysis and assessment regarding applications of ionizing radiation. These conditions were expected to address: (1) the choice of subtopics for the physics module, (2) the ideas about ionizing radiation among lay people, and (3) the resistance to changing these ideas. Such conditions could not be identified by a single study, so a number of studies were planned. These studies are summarized in this chapter. (A full account of the studies is found in Eijkelhof, 1990.)

Choice of Subtopics

A Delphi study was conducted with radiation experts to find out which applications of ionizing radiation ("contexts") and which scientific content would be suitable for inclusion in the unit, given its aim to have students learn to use scientific knowledge in reasoning about radiation risks. We expected that people who are experienced radiation experts and in close contact with lay people in a variety of situations may be of help in making balanced choices among the large number of concepts and applications in this area.

This study was held in three rounds. In each round, at least 35 participants responded to a questionnaire. The results of each round were used to formulate questions for the next round. Participants were carefully selected to represent a variety of fields and opinions on the risks of ionizing radiation.

Lay Ideas About Ionizing Radiation

The investigations into lay ideas about risks and ionizing radiation consisted of an analysis of press reports, questions to the radiation experts in the Delphi study, and interviews with students before they had received any instruction. The lay ideas in the press reports were found in hundreds of clippings about nuclear issues in Dutch and British newspapers and magazines. Although many of those news clippings dealt with the Chernobyl accident, a considerable number of other issues were also covered. The additional questions to the experts in the Delphi study dealt with the nature of lay ideas that the radiation experts found important in their work with lay people. The interviews with 30 students from five different schools (ages 15 and 16) focused on such topics as Chernobyl, food irradiation, health applications, radioactive waste, and background radiation.

Resistance to Change

We studied the resistance by students to changing their preconceived ideas about risk and ionizing radiation. After a series of lessons on ionizing radiation, two groups of students (those who studied the PLON unit and those who studied more conventional teaching materials) responded to a questionnaire with open-ended and multiple-choice items. The 138 pre-university respondents were 17 to 18 years of age and attended five different schools.

RESULTS ON THE CHOICE OF SUBTOPICS

In the first round of the Delphi study, the radiation experts mentioned a large number of applications that could act as contexts for teaching ionizing radiation from a risk perspective. In the second and third rounds, a consensus developed on the relative importance of some of these applications. The experts agreed on four criteria for selecting contexts: (a) the contribution to the "collective dose" (the sum of the dose equivalents received by a specific population), (b) the degree to which members of the public get involved in the applications, (c) the importance of the social implications of the radiation risks, and (d) the variety of applications. The group agreed that the following set of six contexts should be recommended for the new PLON unit:

1. Background radiation from a variety of sources
2. Medical applications (diagnostic and therapeutic)
3. The emission of nuclear substances by power stations (under normal conditions and after an accident)
4. The storage of nuclear waste in various ways
5. Fallout from nuclear weapon explosions
6. Tracers in scientific research and industry

Contexts 1 and 2 relate primarily to criterion (a) *collective dose,* although context 2 also relates significantly to criterion (b) *degree of involvement* because members of the public become most involved at a personal level with medical applications of ionizing radiation. Contexts 3, 4, and 5, are often the object of news reports about accidents and controversies, and therefore are closely related to criterion (c) *social implications*. The final context 6 relates directly to the last criterion (d) *variety of applications*. Tracers in scientific research and industry contribute to the variety of applications of ionizing radiation. Moreover, the experts considered tracers to be an example of a noncontroversial application, and for this reason as well, they felt that tracers should be one of the contexts in the PLON unit.

In view of their six recommended contexts, the participants in the Delphi study agreed upon the scientific content that should be in a unit with a risk perspective. As expected, the radiation experts mentioned a number of concepts already included in most physics curricula (for example, concepts related to the structure of the nucleus, the radioactive decay process, types of radiation, detection, and nuclear fission). One could classify this part of the curriculum as basic knowledge about atomic and nuclear physics.

A second category of scientific content is less familiar to physics education and could be labeled "basic knowledge about radiation protection." According to the radiation experts, this category should include: (1) irradiation, absorption, dose (equivalent), and influence of distance and medium; (2) contamination, and spreading of radioactive substances in the environment and the human body; (3) effects of ionizing radiation, the distinction between early and late effects, those from high and low doses, and somatic and genetic effects; and (4) safety aspects, such as radiation monitoring, safety precautions, radiation norms, and the ALARA-principle.

Many recommendations by the radiation experts corresponded to the content of the original version of the PLON unit, "Ionizing Radiation." Recommendations not included in the original unit were contexts 3 and 6—the emission of radioactive substance and the importance of tracers. In the area of scientific content, the nature of contamination became a new subtopic.

Of course, scientific experts are not the only ones who should decide the content of a curriculum. In the final part of this chapter, I discuss the value and limitations of involving radiation experts in formulating a curriculum's content.

RESULTS FROM THE STUDIES ON LAY IDEAS

Our investigation into lay ideas about ionizing radiation took place in three different domains: articles in newspapers and magazines, responses by radiation experts participating in the Delphi study, and ideas expressed by students. The results from each study are described here.

The Press

Press reports about radiation issues often include terms such as "radiation," "radioactivity," "dose," and "radiation level." This should not be surprising, of course. More interesting is the fact that these reports use the terms differently than do scientists (Eijkelhof & Millar, 1988; Lijnse et al., 1990). One example is "radiation." It is often used in the press when scientists would use the term "radioactive material;" for instance: "Radiation is still pouring into the air from a fire raging at the plant," "The wind is carrying the radiation over Scandinavia," and "Students . . . were contaminated by radiation." Similarly, the term "radioactivity" is often applied by the press to situations in which scientists would use the terms "radiation" or "radioactive material."

The press also conveys the common idea that radiation is, in one way or another, conserved after it has entered an object or person; for example: "Spinach has been exposed to too much radiation," "Radiated crops would not be salable," and "In case of a nuclear disaster our water purifying installations could only remove 30% to 50% of the radiation." Indeed, the lay idea of radiation makes it impossible to make a proper distinction between situations of irradiation and contamination. There is also no distinction in news reports between "activity" and "dose." Both terms, of which the latter is more commonly used, seem to indicate a quantity of radiation contained by food, air, water, or the human body. Both terms are used in combination with safety levels to indicate the degree of danger of a situation.

Radiation Experts

In the Delphi study, the radiation experts described the lay ideas that they often encountered in their professional contacts with the public, discussed the importance of these lay ideas in terms of risk assessment, and helped identify common ideas about the risks of ionizing radiation. The lay ideas most often encountered with the public were similar to those we found in the news reports. Some ideas revealed a lack of insight into the scientific meaning of concepts such as radiation, radioactive material, contamination, irradiation, activity, and dose. Other ideas represented common notions about the risks of radiation, for example, radioactivity and radiation are always dangerous (association with cancer); radiation for medical purposes is less dangerous than radiation used for other purposes; radiation standards indicate a safety level (safe below, dangerous above); and a nuclear power station is as dangerous as a nuclear bomb. According to the participants of the Delphi study, some lay ideas have more serious consequences for risk estimation than others. Not distinguishing between irradiation and contamination, for instance, was illustrated by a number of serious examples from a variety of contexts (Eijkelhof, Klaassen, Lijnse, & Scholte, 1990). These examples showed that people are often worried, either too much or sometimes too little, due to their lay conceptions of radiation and risk.

Students

A third study focused on students' ideas about radioactivity, ionizing radiation, and risk, in the contexts of Chernobyl, medical use of radiation, radioactive waste disposal, food irradiation, and background radiation. Students' ideas within these contexts show that common sense and incoherent bits of information from the media dominate many of the students' views. Scientific notions play a small or nonexistent part in their thinking. Reasoning appears to be centered around the perceived risk of radiation for people. The nature of the effects of radiation is quite well-known and does not differ from context to context, in spite of the perceived seriousness of the radiation hazard in each context.

Students seem to be less concerned with the nature and origin of the radiation. They often make analogies. These analogies are based on the characteristics of the contexts, especially the function of radiation and the saliency of safety measures. Propagation of radiation is only seen as relevant if the source is at a large distance and is considered dangerous (for example, Chernobyl and radioactive waste). When sources are nearby and inside buildings (for example, food irradiation, health applications, and radon from building materials), ventilation is considered to be useful to prevent an accumulation of radiation. The scientific notion of absorption was seldom found.

Students seem to have conservation ideas about radiation, which could be summarized as: When an object (such as food or a wall) receives radiation, the radiation will accumulate in the object; when the amount of radiation is large enough, the object will itself start emitting radiation. Students also appear to attribute meanings to the terms "radiation," "contamination," "radiation standards," and "radioactivity," which differ from the accepted scientific meanings.

Regarding the meaning of radiation, we found that students also associate the idea with other kinds of radiation, especially with light, but also with heat, radio waves, and sound waves. Invisibility is a property of radiation acknowledged by almost all students. Some students seem to reify radiation by calling it a kind of gas ("radon"). A large proportion of students make distinctions among natural radiation, x-rays, and radioactive radiation. Students draw these distinctions partly from their ideas about the differences in the effects of these three kinds of radiation, and partly from the differences in function of these kinds of radiation.

Indiscriminate use of the terms "contamination" and "irradiation" was very common. Many students spoke of contamination when someone or something has received a certain amount of radiation, sometimes specified as "a surplus" of it or "more than normal."

Analysis of the students' answers also showed that students seem to hold different meanings for "radiation standards." We recognized four distinct views:

1. A *regulation* view expressing the idea that standards are what the regulatory bodies consider to be acceptable. This would apply to contamination of food and air, and to irradiation of people.
2. A *threshold* view suggesting that there will be zero risk below a certain standard level. The body is expected to be able to cope with small amounts of radiation.
3. A *probabilistic* view in which the risks are seen as more or less directly proportional to the received dose. Below the standards the risk is generally seen as small.
4. A *distrust* view. Standards are seen as meaningless because all radiation is dangerous.

When using the term "radioactivity," students often meant radiation. Consequently, students spoke about radiating, releasing, or emitting radioactivity. They defined radioactivity as "radiation" or "an accumulation of radiation." We also discovered meanings of radioactivity that meant the source of radiation. Here we refer to students' descriptions, such as "radiation-emitting substances," "particles," "an atom of radioactivity having, handing over, or containing, radioactivity." Very often the effects of radiation ("dangerous" and "noxious") were included in the definition of the term radioactivity.

Finally, in the interviews we identified forms of reasoning that we labeled commonsense reasoning. Students draw conclusions that seem logical to them because the conclusions are based on the uses of radiation, on personal experience, or on the existence of safety measures or protests about safety. Examples of such reasoning are: "If you look at how the workers in a food irradiation plant have to be protected with special clothing, it could not be right for an apple to receive a dose of radiation." "Food irradiation is not dangerous, otherwise they wouldn't do it."

RESULTS ON THE PERSISTENCE OF LAY IDEAS

Physics teachers may recognize the lay ideas reported above, but teachers might expect that these ideas would change as a result of instruction. To test this prediction we designed a questionnaire with a number of lay ideas. It was presented to 138 students who had received several weeks of instruction on the topic of ionization radiation toward the end of a preuniversity physics course. About one-half of the students had studied from the original version of the PLON unit, "Ionizing Radiation," while the others had studied from traditional physics materials. The following eight lay ideas appeared to persist with a majority of students:

1. Radioactive contamination means that someone is contaminated with radiation.
2. The period in which nuclear radiation from an external source remains active in a cancer patient depends on half-life or length of irradiation.
3. Radiation might accumulate in the human body.
4. The term "radioactivity" might be well described as "radioactive radiation."
5. After an accident in a nuclear power station, the radiation might be spread by the wind.
6. Radiation might be stored in food that is irradiated, so it is dangerous to eat that food for some time.
7. X-rays have effects on the human body that are very different from those of nuclear radiation.
8. X-rays should be extracted from air in order to reduce radiation risks.

There was no significant difference in the presence of these lay ideas between the PLON students and the students who worked with traditional books. In both learning materials, no attention was given to the existence of lay ideas. Apparently, teaching the topic of ionizing radiation without paying attention to lay ideas has little effect on students altering those ideas.

Some people may be surprised by the persistence of the lay ideas. In our view, several reasons may contribute to this persistence. These are summarized here.

Many of the lay ideas mentioned in the previous section logically support each other. They are linked in a more or less consistent pattern. Those ideas are part of common culture. They fit in the way people communicate about radiation issues and are reinforced by the media. The lay ideas seem to work well in daily life. People feel that they understand in broad terms what the risks of radiation are about. Moreover, people are not likely to meet situations in which they themselves feel that their knowledge base is inadequate.

These lay ideas are not adequately addressed in science education. They are either ignored, seen as isolated misconceptions, or assumed to disappear after didactic teaching.

The results of the studies described in this chapter indicate that lay ideas are obstacles to acquiring knowledge and skills to reason about the risks of ionizing radiation. Before science education can contribute to public understanding of radiation issues, teachers need to be aware of their students' knowledge and reasoning base (lay ideas). But awareness alone is not sufficient. Specific teaching strategies are needed as well.

A RESEARCH STUDY ON INSTRUCTION

As a result of the research described above, we investigated teaching strategies that might tunnel through the obstacle of lay ideas (Eijkelhof, Van der Valk, & Lijnse, 1992). Our work provided clues on how to improve instruction. For example, we concluded that more emphasis should be given to the differences and communalities between lay and scientific views. Within the scientific view, the meaning of concepts (such as radiation, absorption, and contamination) and the relation among the concepts should get proper attention; for instance, by using the word tasks suggested by Sutton (1992). Students should practice using their scientific knowledge when they consider realistic risk issues. Contexts such as food irradiation, x-rays, cancer treatment, and industrial accidents with open sources appear to be very useful for this purpose.

CONCLUSION

From all of these studies, some lessons may be drawn for STS education as a whole. If one of the aims of STS is for students to learn to use scientific knowledge in realistic situations, it is important to assess what kind of knowledge is most useful in which situations. Just piling concepts into contexts may not be the most adequate way to reach this aim.

In our view, it is necessary to take two important factors into account.

The first factor deals with the relevance of concepts in view of the importance of a set of contexts; that is, the cross-contextual value of concepts should be a point of consideration. From this point of view, concepts such as radiation, absorption, and contamination, may be more relevant than rather detailed knowledge about specific contexts. In that light, debates in science classes about the usefulness of nuclear power stations may deserve a lower priority than given by STS colleagues and ourselves. This argument adds to David Layton's critical comment in Chapter 4 of this book concerning the aim of improving citizen participation in the science-technology issues confronting society.

The second factor addresses the learning difficulties related to the progress from lay ways of thinking toward more scientific forms of reasoning. Teachers and textbooks should take into account the lay ideas of students and should possibly start with these ideas.

Studies similar to those reported here could play an important role in future STS education research. Science and technology experts in a particular field (for example, nutrition, electrochemistry, ecology, or thermodynamics) can identify concepts that are relevant in a variety of everyday contexts (the first factor mentioned above). But these same experts are not likely aware of the tenacious obstacles that keep students from moving away from their lay ways of thinking toward scientific ways of thinking (the second factor). Although science education researchers generally do not have experience with the first factor, they do have insight into the second factor. Both factors are important to future curriculum development and student assessment of any STS education dedicated to teaching students to use scientific knowledge in everyday situations. But both factors need a research base with which STS educators can make informed decisions. This research base requires the cooperation of experts in science and technology and experts in science education who study student learning. As a consequence of such a research base, more curriculum materials could be developed in a way that enhances the quality of STS education.

Chapter 20

Collaborative Research and Development to Produce an STS Course for School Science

Glen Aikenhead

This chapter describes how a practical deliberative enquiry guided the development of an STS science textbook for the high school.

In its education study *Science for Every Student,* the Science Council of Canada (1984) called for a renewal of science education, advising educators to teach scientific concepts and skills embedded in social and technological contexts relevant to all students. Curriculum theorist Joseph Schwab (1974) had argued the need for educators (1) to engage in practical deliberative enquiry (as opposed to theory-generating enquiry), and (2) to pay closer attention to what happens with students in the classroom. Both of Schwab's points inform the present study: a practical deliberative enquiry *with students* to produce an STS science textbook commensurate with recommendations proposed by the Science Council of Canada.

Deliberative enquiry guided the Science Council of Canada's education study and influenced a subsequent science education study undertaken by the Saskatchewan Department of Education. Because both deliberative enquiries define the context of the present study, they are summarized briefly before addressing the present study.

NATIONAL AND PROVINCIAL DELIBERATIONS

The Science Council of Canada put Schwab's deliberative enquiry approach into action during a unique science education study that spanned 5 years, 1979–1984 (Orpwood, 1985). The study ensured that significant problems were identified, that appropriate data were collected, and that these problems and data were considered by the diverse stakeholders (including students) attending 2-day deliberative conferences held across

Canada in 1983. The deliberative conferences unfolded as Schwab (1978) had predicted:

> Deliberation is complex and arduous. . . . It must try to identify the desiderata in the case. It must generate alternative solutions. . . . It must then weigh alternatives and their costs and consequences against one another, and choose, not the *right* alternative, for there is no such thing, but the *best* one. (pp. 318–319)

This "best" solution included the following recommendations (Science Council of Canada, 1984). Along with scientific concepts and skills, students should learn an appreciation for: (1) authentic science—the nature of science and scientists, including the way science generates and uses its knowledge, (2) technology in Canada, and (3) the interrelationships among science, technology, and society.

These recommendations, as well as the process of deliberation itself, greatly influenced a provincial project that was designed to change the science curriculum in Saskatchewan (Hart, 1989). In the Saskatchewan science study, conducted between 1986 and 1987, an overwhelming 92% of Saskatchewan science teachers participating in the deliberative enquiry endorsed the goal of "scientific literacy" defined as a *balance* among seven subgoals:

1. The nature of science itself
2. The key facts, principles, and concepts of science
3. The intellectual processes used in doing science
4. The interactions among science, technology, society, and the environment
5. The values that underlie science
6. The manipulative skills required for doing science
7. Personal interests and attitudes toward scientific and technological matters.

Although 88% of the teachers believed that a science-technology-society-environment (STSE) emphasis to the curriculum should be adopted (the Saskatchewan equivalent of STS), teachers expressed many concerns about: (1) the balance between an STSE emphasis and other emphases, for example, a solid foundation for the next level of science study (Roberts, 1988); (2) the evaluation of students with respect to STSE goals; (3) the availability of appropriate teaching materials; and (4) the need to teach controversial issues. In other words, teachers had positive attitudes but were certainly cautious about changing their science curriculum toward an STSE approach.

The Saskatchewan study had established the goal of "scientific literacy" and a curriculum emphasis on STSE. Saskatchewan science teachers expressed, among other things, a need for classroom materials—such as a textbook—to support them in their efforts to implement a new science curriculum.

THE STUDY: DEFINITION AND PURPOSE

A curriculum exists in three different phases or forms: (1) the *intended* curriculum developed by curriculum specialists or committees authorized by governments; (2) the curriculum as *taught* by classroom teachers, the translation of the intended curriculum by teachers when they prepare and teach lessons; and (3) the curriculum as *learned* by students. Ideally, the intended, taught, and learned curricula should be similar. In reality, however, they differ widely (Cronin-Jones, 1991).

The similarity among the intended, taught, and learned curricula may be strengthened by using classroom materials—for instance, a textbook—designed to give students experiences that clearly convey the intended curriculum (Aikenhead, Chapter 16). A textbook influences the content of the taught and learned curricula whenever the textbook is a teacher's primary resource, a condition found in most North American high school science classes (Gallagher, 1991).

In Saskatchewan's deliberative meetings, curriculum developers and classroom instructors discussed what ought to be taught in science classes. Although the perceived needs and interests of students were given high priority, students were not involved in the provincial process of curriculum deliberation.

However, students can play a significant role in determining the quality of the *learned* curriculum (Aikenhead, 1982; Eijkelhof, Chapter 19; Kortland, 1992; Solomon, 1983). This role was explored in the present study by involving students in a new type of deliberative process: a collaboration with textbook authors to produce classroom materials that strengthen the similarity among the intended, the taught, and the learned curricula.

In the study, the textbook developer (Glen Aikenhead) consulted and negotiated with students and teachers. Students worked through, and deliberated over, draft versions of the textbook. Students offered concrete guidance on how to achieve the intended curriculum. As described below in the Results section, students contributed significantly to the textbook's content, structure, and language.

The study was primarily concerned with students learning a curriculum that had been rationalized by two successive Schwabian deliberations.

The study, however, went beyond practical enquiry characterized by defensible decision making. The study included research and development that led to the publication of an academic STS textbook and teacher's guide, *Logical Reasoning in Science & Technology,* LoRST (Aikenhead, 1991a, 1991b). The study not only focused on the classroom experiences of students (the learned curriculum), but also collaborated with those students and their teachers in a type of practical enquiry to produce classroom materials that were: (1) in harmony with the intended curriculum, (2) usable by teachers with limited inservice training, and (3) consistent with students' views on relevancy and practical appropriateness. This chapter describes this deliberative research and development process.

Overview of the Study

The research and development of LoRST followed a three-stage sequence that took advantage of the classroom realism well-known to teachers and students. These phases are summarized here. In phase 1 (an 8-week project), I wrote and taught draft 1 in a local high school in 1987. Based on this classroom collaboration with students, the text was modified and expanded to yield draft 2. By initiating the project in a classroom setting: (1) the classroom materials evolved on the spot with average grade 10 students, (2) appropriate teaching strategies were identified (Aikenhead, 1988b), and (3) a rough draft of the teacher's guide was written. This second draft of LoRST was used in 1988 (phase 2, a 20-week project) by three volunteer teachers who received no inservice training. The teachers taught in three different schools, representing a full cross section of student background and abilities. Classes were observed daily as teachers used draft 2 of the text and teacher's guide. This collaboration with students and teachers led to another revision of the student materials and teaching strategies. As a result of this closely monitored project, LoRST was polished into draft 3. In 1989–1990 (phase 3), draft 3 was field tested across Saskatchewan and evaluated by 30 teachers both sympathetic to, and critical of, an STS approach to teaching science. Teacher feedback resulted in further revisions to LoRST. The resulting material (student text and teacher's guide) was published in 1991 and adopted by Saskatchewan as a principal textbook for grade 10. Two other provinces adopted LoRST in 1993.

The former grade 10 syllabus in Saskatchewan had been a traditional physical science course with an emphasis on chemistry. The LoRST project was initiated with the assumption that a similar content would be the basis of the new curriculum (a false assumption, as it turned out).

The three-phase research and development process that produced LoRST will be much easier to understand if one has an idea of the product of the study—the textbook.

Product of the Study

LoRST teaches scientific content in conjunction with critical reasoning skills to a target audience of grade 10 students of average (or above average) academic ability. (For a detailed description of LoRST see Aikenhead, 1992b.) Students learn scientific facts, concepts, and principles from physics, chemistry, and biology in a way that connects those facts, concepts, and principles with the students' everyday world. The interdisciplinary nature of LoRST places it in category 5 of the STS science scheme presented as Table 5.1 in Chapter 5.

The textbook begins with courtroom testimony by scientific experts—a social context familiar to students. This creates the need to know a host of science concepts and logical reasoning skills. In LoRST, the social issue of drinking and driving creates the need to know (1) the technology of the breathalyzer; (2) how science and technology interact with each other, and how they both interact with various aspects of society; and (3) scientific content such as mixtures, concentration, chemical reactions, photometry, electrical circuits, and the biology of body cells and systems. Although the content is "driven by" the social issue of drinking and driving, the content is not limited to that social issue. For instance, students solve concentration problems in the world of recipes, false advertising, toxic chemicals, and farm fertilizers. Classification of mixtures is introduced in the context of the Red Cross and is developed via the technology of salad dressings. Electricity concepts are learned in order to bridge the gap between atomic theory and the household appliances familiar to adolescents (both female and male). Heat and temperature are taught in an historical context, accompanied by inquiry labs requiring students to construct relevant concepts.

LoRST's emphasis on logical reasoning reflects the mandate to improve students' critical thinking skills (Aikenhead, 1990; Byrne & Johnstone, 1978; National Science Foundation, [NSF] 1990). Specific critical reasoning skills are taught in Unit 3, "Science & Critical Thinking: The Logic Game." These skills are then applied throughout the book. More important than the individual reasoning skills themselves is the increase in students' *predisposition* (habits of the mind) to analyze, to question, and to articulate a reasoned argument (McPeck, 1981).

RESULTS

This section describes specific instances of collaboration in which students and teachers contributed to the content, structure, and language of the final product—*Logical Reasoning in Science & Technology*. Although each page in the textbook has a unique story to tell about how it evolved over the three phases of the study, only a few typical examples are described here. (More details are provided by Aikenhead, 1991c.) The examples address four topics: selection of the content, structuring the content, students "editing" the language of the textbook, and lastly, contributions by teachers.

Selection of Content

The social issue of drinking and driving was selected as an organizing STS theme in LoRST. The issue is of critical importance to every community. Alcohol is the most fatal toxic chemical in the environment when it pollutes a driver's body. The issue of drinking and driving also demands particularly realistic decisions by students, rather than the more idealistic, hypothetical decisions sometimes associated with global issues (Carter, 1991). Students in grade 10 know that they will soon take personal action based on their own decisions about alcohol and driving. Thus, an action orientation to decision making is particularly realistic for students (Rubba, 1991). Questionnaire feedback from 91 students participating in phase 2 indicated that most (83%) found the issue of alcohol and driving to be relevant and enjoyable, though they would have preferred studying science via the issues of sex and drugs.

During phase 1 of the research and development (R & D) study, students vigorously quizzed me about facts and principles related to how alcohol would affect their bodies. This was science content not found in the old curriculum and therefore not intended for LoRST. (It is now part of the new curriculum.) Students' interest and curiosity convinced me to compose a new unit, "How Alcohol Affects the Body." It was added to the end of LoRST version 1. I designed this new unit in a way that systematically addressed students' questions, but also applied previously learned physical science content to biological systems; for instance, concepts such as diffusion, chemical changes, and parallel and series circuits.

To provide practice at applying physical science content to the everyday world, some decision-making activities were developed in version 2. Students fell short of my expectations of how easily they could reach a decision. When I clarified these expectations (the intended curriculum) during a class discussion, students pointed out that more guidance was needed to lead them through what they perceived to be unfamiliar territory. In other words, students suggested how the learned curriculum could be more consistent with the intended curriculum. As a consequence, another unit, "Decision Making," was tacked on to the end of LoRST version

2. It described a 10-step sequence to follow when making a decision (Aikenhead, 1985a).

Not only did students initiate the development of these two new units, they contributed to the content of other units. In a section on electricity, for instance, students were the source of questions that now appear in the text. During phase 1 of the study, I overheard a conversation between two girls about accidentally plugging a hair dryer into a 220 volt source. "It killed my hair dryer," one said. Version 2 of LoRST incorporated her story into a question that asked why 200 volts would "kill" a hair dryer.

Structuring the Content

Three examples of structuring the content in LoRST will further clarify the deliberative process that characterized the research and development study.

A postlab discussion of an activity called "Experimenting with Mixtures" provided a forum for identifying the activity's successes and difficulties. Students expressed frustration over my expectation that they could inductively derive a distinction between homogeneous and heterogeneous mixtures. Students' preconceptions of mixtures were far richer and more complex than the scientific dichotomy of homogeneous/heterogeneous. "Why couldn't we just read about it before the activity?" they asked. The section was revised accordingly.

The second example of structuring content in LoRST deals with a section on electricity. In version 1, students were given the challenging task to explore electricity ("bulbs and batteries" revisited) and to write what they discovered. This approach was fashioned after the Nuffield science projects in the United Kingdom, in which students learned almost all their science content from lab activities. The expectation of the LoRST research and development project, however, was to turn the students' work produced during version 1 into an activity for version 2. With appropriate editing, version 2 incorporated the students' work into a single activity comprised of five interconnecting parts that took about 5 hours to complete. The students who studied version 2 did not like this long and involved activity. Instead, they wanted an activity to be more focused on one particular topic. Therefore, the five-part activity of version 2 was restructured into four activities embedded in didactic text that introduced or reinforced the four activities. In doing so, the topic of electricity was reorganized around two themes: scientific theory and scientific law. These themes reinforced the textbook's epistemic distinction between theories and laws.

The third illustration of students structuring text content occurred when one of LoRST's activities failed. I wanted students to analyze a current article from the *New England Journal of Medicine* to see a 1990 example of the tentativeness of science, and learn more about scientists participating in consensus making. I wrote a synopsis of the article. The first time students read it, they could not decide whether the conclusion reached was logical or not. (The conclusion was somewhat controversial within the scientific community.) During a classroom discussion of the article, I acknowledged the impasse the class had come to, and wondered aloud how the impasse might be resolved. One student pointed out that there were actually two main conclusions to the article, not one, and that one conclusion was logical while the other one was not. Her analysis was written into the next version of LoRST. Now students are given two conclusions and asked to comment on the logic of each conclusion. Difficult science content can sometimes be clarified by students themselves, and can then be restructured to conform with an adolescent logical perspective.

Teachers were often helpful at identifying student problems and at articulating student concerns. Throughout phase 2, students' problems and preconceptions led me to reformulate the content and to restructure the text. In other words, collaboration with students and teachers identified problems, generated alternatives, and made it possible to choose the best solutions. A Schwabian type of practical enquiry was working.

Students "Editing" the Textbook

Students will gladly express confusion or frustration over something they do not understand. By sitting in their classroom day in and day out, one can detect not only their confusions or frustrations, but also how they spontaneously clarify or correct the problems. Typical examples will illustrate the point.

Lab instructions were modified when I observed students making errors or making impromptu improvements to the instructions. The intent of LoRST was to have students *reason* with scientific ideas rather than simply memorize them. Consequently, the LoRST lab instructions were written much differently than lab instructions designed to have students verify facts. Teachers' extemporaneous modifications to labs (in phase 2) also contributed to the clarity and efficiency of LoRST's lab instructions.

Another issue (related to the language in LoRST) deals with the text's writing style. Young, Ruck, and Crocker (1991, p. 46) claim, "Science texts violate students' expectations because the language is unlike anything they have previously encountered. The way science books present ideas is a discrepant event in the students' experience." The traditionally formal, succinctly dense, science language makes textbooks science-centered, not student-centered. Rather than requiring students to completely change their reading style in order to understand the science text language, one can modify the text's language to conform more with student expectations. This has been achieved in LoRST. The text is unusually narrative (sometimes even chatty), filled with visual imagery, and structured from a student's perspective. Thus, students indirectly influenced the writing style of LoRST because of the student-centered approach to the use of language.

However, students had a direct effect on the writing style as well. Occasionally, the text is literally interrupted by a comment from a student, usually at some key point in the development of an idea. For instance, the water content of body tissue is studied in LoRST. The textbook points out that a 70-kg female athlete (high muscle content) would be less affected by an alcoholic beverage than a 70-kg potbellied male (high fat content), all other variables being equal. Then a cartoon-like balloon interrupts the paragraph and points to a silhouette of a student's head in the margin. Inside the balloon the following passage appears: "That can't be true," he protested. "I read in a book that women are more affected by alcohol than men" (Aikenhead, 1991a, p. 223). The text then responds in a dialogue fashion and introduces students to the idea of a statistical fact, the type to which the student was referring. This dialogue between student and textbook paraphrases an actual classroom interaction. The dialogue emphasizes a discrepancy between a student's preconception about scientific facts (they are either true or false) and the curriculum's conception of scientific facts (they are context dependent and probabilistic). This style of highlighting discrepancies between the intended curriculum and students' preconceptions was inadvertently suggested to me by the poignancy of certain key student-teacher exchanges that took place in the classroom. These exchanges are recaptured in LoRST as a dialogue between the student and the textbook. The practical enquiry of collaborating with students allowed me to build on students' natural reaction to classroom events, and to judiciously incorporate some of those reactions into the text.

The informal, unorthodox, narrative writing style in LoRST received strong endorsement from the 91 students surveyed in phase 2. Ninety-two percent thought the style made the textbook easier to understand, while 5% thought it made it more difficult.

Teacher Contributions

It is evident from the examples cited above how the three teachers in phase 2 of the study contributed to the development of LoRST. Their contributions normally arose spontaneously during classroom action, where idiosyncrasies and constraints abound. Helped only by a rough draft of the teacher's guide for LoRST, these teachers explored new territory of STS instruction. By observing each lesson (the taught curriculum) and by listening to a teacher's out of class comments, I modified the teacher's guide in several ways: by clarifying directions, by adding specific teaching suggestions, and by writing background information. Day by day the teachers also identified problems for me to solve in the student text.

The revision of LoRST at the end of phase 2 would certainly have been suitable for most grade 10 students. After all, students themselves had contributed to its content, structure, and language. But would LoRST work with other teachers? Critics of STS claim it would not (Walberg, 1991).

Phase 3 of the study, a provincewide field test of manuscript version 3, involved 30 teachers chosen by Saskatchewan Education, some of whom were supportive but others who were skeptical or even critical of an STS approach. Phase 3 was not an investigation into teacher implementation of LoRST nor an evaluation of LoRST. Rather, it was a practical enquiry into developing LoRST further. A few examples of collaboration in phase 3 will illustrate how LoRST was improved.

Unlike the day-to-day collaboration with students and teachers in phases 1 and 2, in phase 3 teachers only identified problems. (Financial and human resources were limited.) The solutions to these problems were left for me to engineer. A short case study about the topic of heat will illustrate typically what happened in phase 3.

A workshop had been organized to help teachers with some new content in LoRST—the explicit treatment of critical thinking in a unit called "The Logic Game." The teacher's guide suggested that students read, analyze, and discuss. This logic game unit had been very successful in another project with average students (Aikenhead, 1979a). That success had been attributed to the fact that the logic content that students read, analyzed, and discussed, did not come from science but from the everyday world of adolescent experience. At the workshop, however, one teacher emphatically explained his class's negative reaction to this logic content, "When my students come into science class, they want to light bunsen burners and get the right answers." Because the teacher was conscientiously implementing the new curriculum, and because he likely represented a prevailing view among science teachers, his comment was taken seriously. Thus, my objective was to design:

1. Activities in which students light bunsen burners and get the right answers.
2. Activities that correlate with the content in "The Logic Game" unit, but could also function independently of this content in case teachers do not teach the logic unit.

3. Activities that reflect a constructivist and STS perspective.
4. Activities that correspond to the emerging curriculum.

One new activity, "The Law of Heating (and Cooling) Bodies," was designed according to the four specifications stated above. The activity extrapolates from the student's everyday world of temperature changes in syrup, cooking oil, and water, to the realm of scientific thinking about specific heat capacities. Students, acting as research teams, explore the variables that seem to affect heat transfer. (Exploration rather than verification occurs.) *Bunsen burners are lit!* Each research team works on one of the variables. Then all the teams get together at an "international conference" where the class reaches a consensus on what to believe about which variables affect heat transfer. To prepare for their conference, students analyze the heat data. Students must reason on the basis of the assumption: "One gram of substance rising one degree Celsius takes in a certain number of joules of heat." After reaching a consensus, students are introduced to the term "specific heat capacity." Next, the heat transfer equation is presented as a mathematical wording to "The Law of Heating (and Cooling) Bodies." Finally, several math problems are solved to gain a facility at using the heat transfer equation. *Students get the right answers!* The teacher who wanted students to light bunsen burners and get the right answers had his specifications met.

Two further activities on heat were developed and cycled through the three phases of the deliberative enquiry process to complete a systematic treatment of heat and temperature (Aikenhead, 1992a). The phase 3 collaboration with the 30 teachers led to the production of several other activities that would not have otherwise been developed. The collaboration also resulted in LoRST looking more like a traditional science text (for example, chapter titles were changed to conform to traditional expectations), but without compromising the harmony between the intended and learned curricula achieved in phases 1 and 2. None of the changes undermined the book's relevancy and practicality to students.

CONCLUSIONS

The ultimate purpose of the LoRST research and development (R & D) project was to strengthen the cohesion between (1) the intended curriculum, advocated by the Science Council of Canada and defined by Saskatchewan's Department of Education, and (2) the curriculum learned by students. Accordingly, LoRST was developed to help teachers translate (modulate) the intended curriculum into a taught curriculum, in a way that more accurately portrays the intended curriculum (Roberts, 1980).

The practical enquiry demonstrated that students can contribute significantly to a textbook's content, structure, and language. By engaging students in tasks in the natural setting of their classroom, an author can attend to information that spontaneously emerges during instruction or to information that thoughtfully evolves from informal discussions with students. Students were most helpful in (1) reflecting on how well the tasks met the provincial curriculum's objectives (that is, the consistency between the learned and intended curricula); and (2) suggesting, indirectly or directly, how to improve on the text material. In a somewhat similar fashion, Kortland's (1992) environmental education project also demonstrates how student interviews and classroom observations can judiciously guide the development of instructional materials.

Further research is needed to shed more light on the learned curriculum that results from using STS materials such as LoRST. One promising avenue of research is the systematic identification of classroom events that appear to have an impact on students' reconceptionalizations of science and STS content. Case studies of a small number of students can clarify the classroom events that affect their understanding of, for instance, the nature of heat, the role of assumptions in scientific thinking, or the nature of scientific decision making. (See, for example, studies by Larochelle and Désautels, 1991; Roth and Roychoudhury, 1993; and Shapiro, 1989.) A case study of students learning science can portray the different ways in which students' feelings, worldviews, and preconceptions interact with the teacher's taught curriculum, fellow students, and the classroom events supplied by a textbook. These interactions provide rich data for better understanding the learned curriculum of an STS science course.

References

Addinell, S., & Solomon, J. (1983). *Science in a social context: Teacher's guide*. Oxford: Basil Blackwell.

Aikenhead, G. S. (1973). The measurement of high school students' knowledge about science and scientists. *Science Education, 57*(4), 539–549.

Aikenhead, G. S. (1974). Course evaluation: 2. Interpretation of student performance on evaluative tests. *Journal of Research in Science Teaching, 11*(1), 23–30.

Aikenhead, G. S. (1979a). Science: A way of knowing. *Science Teacher, 46*(6), 23–25.

Aikenhead, G. S. (1979b). Using qualitative data in formative evaluation. *Alberta Journal of Educational Research, 25*(2), 117–129.

Aikenhead, G. S. (1982). A retrospective account of the development of a novel curriculum in science: Prospects for change. In R. Butt, J. Olson, & J. Daignault (Eds.), *Insiders' realities, outsiders' dreams: Prospects for curriculum change* (pp. 124–138). Vancouver, BC: University of British Columbia, Centre for the Study of Curriculum and Instruction.

Aikenhead, G. S. (1985a). Collective decision making in the social context of science. *Science Education, 69*(4), 453–475.

Aikenhead, G. S. (1985b). Science curricula and preparation for social responsibility. In R. W. Bybee (Ed.), *Science-technology-society*. 1985 NSTA Yearbook. Washington, DC: National Science Teachers Association.

Aikenhead, G. S. (1986). The content of STS education. *STS Research Network Missive, 2*(3), 18–23.

Aikenhead, G. S. (1988a). An analysis of four ways of assessing student beliefs about STS topics. *Journal of Research in Science Teaching, 25*(8), 607–629.

Aikenhead, G. S. (1988b). *Teaching science through a science-technology-society-environment approach: An instruction guide*. Regina, Saskatchewan: University of Regina, SIDRU, Faculty of Education.

Aikenhead, G. S. (1990). Scientific/technological literacy, critical thinking, classroom practice. In S. Norris & L. Phillips (Eds.), *Foundations of literacy policy in Canada* (127–145). Calgary, Alberta: Detselig.

Aikenhead, G. S. (1991a). *Logical reasoning in science & technology* (Student Text). Toronto, Ontario: John Wiley of Canada.

Aikenhead, G. S. (1991b). *Logical reasoning in science & technology* (Teacher's Guide). Toronto, Ontario: John Wiley of Canada.

Aikenhead, G. S. (1991c, June). *Transposing STS science curriculum policy into a high school textbook*. A paper presented to the annual meeting of the Canadian Society for the Study of Education, Queen's University, Kingston, Ontario, Canada.

Aikenhead, G. S. (1992a). How to teach the epistemology and sociology of science in a historical context. In S. Hills (Ed.), *The history and philosophy of science in science education* (Vol. 1, pp. 23–34). Kingston, Ontario, Canada: Queen's University, Faculty of Education.

Aikenhead, G. S. (1992b). Logical reasoning in science and technology. *Bulletin of Science, Technology & Society, 12*(3), 149–159.

Aikenhead, G. S. (1992c). The integration of STS into science education. *Theory into Practice, 31*(1), 27–35.

Aikenhead, G. S., & Fleming, R. W. (1975). *Science: A way of knowing.* Saskatoon, Saskatchewan, Canada: University of Saskatchewan, Department of Curriculum Studies.

Aikenhead, G. S., Fleming, R. W., & Ryan, A. G. (1987). High school graduates' beliefs about science-technology-society: 1. Methods and issues in monitoring students views. *Science Education, 71*(2), 145–161.

Aikenhead, G. S., & Ryan, A. G. (1989). *The development of a multiple-choice instrument for monitoring views on science-technology-society topics.* Ottawa, Ontario: Social Sciences and Humanities Research Council of Canada.

Aikenhead, G. S., & Ryan, A. G. (1992). The development of a new instrument: "Views on science-technology-society" (VOSTS). *Science Education, 76*(5), 477–491.

Aikenhead, G. S., Ryan, A. G., & Désautels, J. (1989, April). *Monitoring student views on science-technology-society topics: The development of multiple-choice items.* Paper presented to the annual meeting of the National Association for Research in Science Teaching, San Francisco.

Aikenhead, G. S., Ryan, A., & Fleming, R. (1989). *Views on science-technology-society* (form CDN.mc.5). Saskatoon, Saskatchewan, Canada: University of Saskatchewan, Department of Curriculum Studies.

Amara, J. M. (1987). Indigenous technology of Sierra Leone and the science education of girls. *International Journal of Science Education, 9*(3), 317–324.

American Association for the Advancement of Science (AAAS). (1989). *Science for all Americans.* Washington, DC: Author.

American Chemical Society (ACS). (1988a). *ChemCom: Chemistry in the community.* Dubuque, IA: Kendall/Hunt.

American Chemical Society (ACS). (1988b). *ChemCom: Chemistry in the community. Teacher's guide.* Dubuque, IA: Kendall/Hunt.

Anyon, J. (1979). Ideology and United States history textbooks. *Harvard Educational Review, 49*(3), 361–386.

Arditti, R. (1979). Feminism and science. In R. Arditti, P. Brennan, & S. Cavrak (Eds.), *Science and liberation* (pp. 350–368). Boston: South End Press.

Arnstein, S. R. (1977). A ladder of citizen participation. In G. Boyle, D. Elliot, & R. Roy (Eds.), *The politics of technology* (pp. 238–245). London: Longman (in association with The Open University Press).

Association for Science Education. (1981). *Science in society: Teacher's guide.* London: Heinemann Educational Books.

Association for Science Education. (1990). *Gender and science education.* Hatfield, Hertfordshire, UK: Author.

Association for Science Education. (1991). *Policy statement.* Primary science. Hatfield, Hertfordshire, UK: Author.

Attfield, R. (1983). *The ethics of environmental concern.* Oxford: Basil Blackwell.

Bacon, F. (1605). *The advancement of learning.* London: Dent & Sons.

Banerjee, A. C., & Yager, R. E. (1992). Improvement in students perceptions of their science teachers, the nature of science, and science careers with science-technology-society approaches. In R. E. Yager (Ed.), *The status of STS: Reform efforts around the world* (pp. 102–109). ICASE 1992 Yearbook. Knapp Hill, South Harting, Petersfield UK: International Council of Associations for Science Education.

Barnes, D., & Todd, F. (1977). *Communication and learning in small groups.* London: Routledge & Kegan Paul.

Benjamin, H. (1939). *The saber-tooth curriculum.* New York: McGraw-Hill.

Benjamin, H. (1948). The saber-tooth curriculum. In D. Harl & C. Drieder (Eds.), *American public education* (pp. 53–57). New York: John Wiley & Sons.

Bernstein, R. (1991). *The new constellation.* Cambridge: Polity Press

Binadja, A. (1992). Development of science process skills when science is taught with a focus on science-technology-society. In R. E. Yager (Ed.), *The status of STS: Reform efforts around the world* (pp. 97–101). ICASE 1992 Yearbook. Knapp Hill, South Harting, Petersfield UK: International Council of Associations for Science Education.

Blandow, D., & Dyrenfurth, M. (Eds.). (1992). *Technological literacy, competence and innovation in human resource development. Proceedings of the first international conference on technology education.* Columbia, MO: Applied Expertise Associates.

Bloch, E. (1986). Scientific and technological literacy: The need and the challenge. *Bulletin of Science, Technology & Society, 6*(2 & 3), 138–145.

Bondi, H. (1985). Society's view of science. In G. B. Harrison (Ed.), *World trends in science and technology education* (pp. 10–13). Nottingham, UK: Trent Polytechnic.

Booth, A., & Jacobs, H. (1990). Ties that bind: Native American beliefs as a foundation for environmental consciousness. *Environmental Ethics, 12*(1), 27–34.

Boston Women's Health Collective. (1971). *Our bodies, ourselves.* New York: Random House.

Breakwell, G. (1990, March). *Young people's attitudes to scientific change.* Paper presented at the Public Understanding of Science conference, London Science Museum.

Bredderman, T. (1982). Activity science—the evidence shows it matters. *Science and Children, 20*(1), 39–41.

Bredderman, T. (1983). Effects of activity based elementary science on student outcomes: A quantitative synthesis. *Review of Educational Research, 53*(4), 499–518.

Bruffee, K. A. (1992). Science in a postmodern world. *Change, 24*(6), 18–25.

Bruner, J. (1966). *Toward a theory of instruction.* Cambridge, MA: Harvard University Press.

Buseri, J. C. (1987). The influence of culture on pupils' questioning habits in Nigerian secondary schools. *International Journal of Science Education, 9,* 579–584.

Bybee, R. W. (Ed.) (1985a). *Science-technology-society.* 1985 NSTA Yearbook. Washington, DC: National Science Teachers Association.

Bybee, R. W. (1985b). The Sisyphean question in science education: What should the scientifically and technologically literate person know, value and do—as a citizen? In Bybee, R. W. (Ed.), *Science-technology-society* (pp. 79–93). 1985 NSTA Yearbook. Washington, DC: National Science Teachers Association.

Bybee, R. W. (1987). Science education and the science-technology-society (STS) theme. *Science Education, 71*(5), 667–683.

Bybee, R. W., Harms, N., Ward, B., & Yager, R. (1980). Science, society, and science education. *Science Education, 64*(3), 377–395.

Bybee, R. W., & Mau, T. (1986). Science and technology related to global problems: An international survey of science educators. *Journal of Research in Science Teaching, 23*(7), 599–618.

Byrne, M. S., & Johnstone, A. H. (1978). Critical thinking and science education. *Studies in Higher Education, 12*(3), 325–339.

Byrne, M. S., & Johnstone, A. H. (1988). How to make science relevant. *School Science Review, 70*(251), 43–46.

Calderhead, J. (Ed.) (1982). *Exploring teachers' thinking.* London: Cassell.

Cannon, S. F. (1978). *Science in culture: The early Victorian period.* New York: Science History Publications.

Carlsen, W. S., Kelly, G. J., & Cunningham, C. M. (1992, March). *ChemCom in sociocultural context: 3. Interdisciplinary analysis of a science-technology-society curriculum.* Paper presented at the annual meeting of the National Association for Research in Science Teaching, Boston, MA.

Carson, R. (1962). *Silent spring.* Boston: Houghton Mifflin.

Carter, C. (1991). STS and access to scientific knowledge. *Theory into Practice, 30*(4), 273–279.

Chamberlain, P. J. (1985). Integrated science as a preparation for 'A' level physics, chemistry and biology. *Research in Science and Technological Education, 3*(2), 153–158.

Chambers, D. W., & Turnbull, D. (1989). Science worlds: An integrated approach to social studies of science teaching. *Social Studies of Science, 19*(2), 155–179.

Champagne, A. B., & Klopfer, L. E. (1982). A causal model of students' achievement in a college physics course. *Journal of Research in Science Teaching, 19*(4), 299–309.

Champagne, A. B., & Newell, S. T. (1992). Directions for research and development: Alternative methods of assessing scientific literacy. *Journal of Research in Science Teaching, 29*(8), 841–860.

Cheek, D. W. (1992a). Evaluating learning in STS education. *Theory into Practice, 31*(1), 64–72.

Cheek, D. W. (1992b). Experiencing the culture of science and technology through adolescent investigations of local STS issues. In S. Hills (Ed.), *The history and philosophy of science in science education* (Vol. 1). Kingston, Ontario, Canada: Queen's University, Faculty of Education.

Cheek, D. W. (1992c). *Thinking constructively about science, technology and society education.* Albany: State University of New York Press.

Chemical Education for Public Understanding Program. (1991). *What is CEPUP?* Berkeley, CA: Lawrence Hall of Science.

Chiappetta, E. L., Sethna, G. H., & Fillman, D. A. (1991). A quantitative analysis of high school chemistry textbooks for scientific literacy themes and expository learning aids. *Journal of Research in Science Teaching, 28*(10), 939–951.

Cohen, I. B. (1960). *The birth of a new physics.* Garden City, NY: Anchor Books.

Collingridge, D. (1989). Incremental decision making in technological innovations: What role for science? *Science, Technology, & Human Values, 14*(2), 141–162.

Collins, H. M. (1985). *Changing order: Replication and induction in scientific practice.* London: Sage.

Collins, J., & Smithers, A. (1984). Person orientation and science choice. *European Journal of Science Education, 6*(1), 55–65.

Comber, L., & Keeves, J. (1973). *Science education in nineteen countries.* Stockholm: Almquist & Wiksell.

Commission of the European Communities. (1977). *Science and European public opinion.* Brussels: Commission of the European Communities.

Cozzens, S. (1990). The disappearing disciplines of STS. *Bulletin of Science, Technology & Society, 10*(1), 1–5.

Cronin-Jones, L. L. (1991). Science teacher beliefs and their influence on curriculum implementation: Two case studies. *Journal of Research in Science Teaching, 28*(3), 235–250.

Cummings, W. K., Gopinathan, S., & Tomoda, Y. (Eds.). (1988). *The revival of values education in Asia and the West.* Oxford: Pergamon Press.

Cunningham, C. M., Carlsen, W. S., & Kelly, G. J. (1992, March). *ChemCom in sociocultural context: 1. General curriculum analysis.* Paper presented at the annual meeting of the National Association for Research in Science Teaching, Boston.

Dampier, W. C. (1948). *A history of science.* New York: Cambridge University Press.

Dart, F. E., & Pradham, P. L. (1976). Cross cultural teaching in science. *Science, 155,* 649–656.

De Vore, P. W. (1992). Technological literacy and social purpose. *Theory into Practice, 31*(1), 59–63.

Dickson, D. (1984). *The new politics of science.* New York: Pantheon.

Doyle, W., & Ponder, G. (1977). The practicality ethic in teacher decision making. *Interchange, 8*(3), 1–12.

Driver, R., & Oldham, V. (1986). A constructivism approach to curriculum development in science. *Studies in Science Education, 66,* 623–633.

Durant, J. R., Evans, G. A., & Thomas, G. P. (1989). The public understanding of science. *Nature, 340,* 11–14.

Durbin, P. T. (1991). Defining STS: Can we reach consensus? *Bulletin of Science, Technology & Society, 11*(4, 5), 187–190.

Dyrenfurth, M. J., & Kozak, M. R. (Eds.). (1991). *Technological literacy.* 40th Yearbook of the Council on Technology Teacher Education. Peoria, IL: Macmillan/McGraw-Hill.

Ehrenreich, B., & English, D. (1978). *For her own good: 150 years of the experts' advice to women.* Garden City, NY: Doubleday.

Eijkelhof, H. M. C. (1986). Dealing with acceptable risk in science education: The case of ionizing radiation. In M. J. Frazer & A. Kornhauser (Eds.), *Ethics and social responsibility in science education* (pp. 189–199). Oxford: Pergamon Press.

Eijkelhof, H. M. C. (1990). *Radiation and risk in physics education.* Utrecht, The Netherlands: Utrecht University, Centre for Science and Mathematics Education.

Eijkelhof, H. M. C., Klaassen, C. W. J. M., Lijnse, P. L., & Scholte, R. L. J. (1990). Perceived incidence and importance of lay-ideas on ionizing radiation: Results of a Delphi-study among radiation-experts. *Science Education, 74*(2), 183–195.

Eijkelhof, H. M. C., & Kortland, K. (1987). Physics in its personal, social and scientific context. *Bulletin of Science, Technology & Society,* 7(1/2), 125–136.

Eijkelhof, H. M. C., & Kortland, K. (1988). Broadening the aims of physics education. In P. J. Fensham (Ed.), *Development and dilemmas in science education* (pp. 282–305). New York: Falmer Press.

Eijkelhof, H. M. C., & Lijnse, P. (1988). The role of research and development to improve STS education: Experiences from the PLON project. *International Journal of Science Education, 10*(4), 464–474.

Eijkelhof, H. M. C., & Millar, R. (1988). Reading about Chernobyl: The public understanding of radiation and radioactivity. *School Science Review, 70*(251), 35–41.

Eijkelhof, H. M. C., Van der Valk, A. E., & Lijnse, P. L. (1992, June). *The effects of a new teaching strategy on recognizing lay-ideas on ionizing radiation and the ability to make risk assessments.* Paper presented to the European Conference on Educational Research, Enschede, The Netherlands.

Eiser, J. R. (1986). *Social psychology.* Cambridge: Cambridge University Press.

Elkana, Y., & Mendelsohn, E. (1981). *Sciences and cultures, anthropological and historical studies of the sciences.* Sociology of the Sciences Yearbook, Vol. 5. Boston: Dordrecht.

Erickson, G., & Erickson, L. (1984). Females and science achievement: Evidence explanation and implications. *Science Education, 68*(1), 63–89.

Fee, E. (1980). Nineteenth-century craniology: The study of the female skull. *Bulletin of the History of Medicine, 53,* 415–433.

Fee, E. (1981, January). *Is feminism a threat to scientific objectivity?* Paper presented at the annual meeting of the American Association for the Advancement of Science, Toronto.

Fensham, P. J. (1988). Approaches to the teaching of STS in science education. *International Journal of Science Education, 10*(4), 346–356.

Fensham, P. J. (1992). Science and technology. In P. W. Jackson (Ed.), *Handbook of research on curriculum* (pp. 789–829). New York: Macmillan.

Finley, F., Lawrenz, F., & Heller, P. (1992). A summary of research in science education—1990. *Science Education, 76*(3), 239–338.

Fishbein, M. (1975). *Beliefs, attitudes, intentions and behaviour: An introduction to theory and research.* Reading, MA: Addison-Wesley.

Fleming, R. W. (1986). Adolescent reasoning in socio-scientific issues (Pt. 1). *Journal of Research in Science Teaching, 23*(8), 677–688.

Fleming, R. W. (1989). Literacy for a technological age. *Science Education, 73*(4), 391–404.

Fourez, G. (1989). Scientific literacy, societal choices, and ideologies. In A. B. Champagne, B. E. Lovitts, & B. J. Calinger (Eds.), *Scientific literacy* (pp. 89–108). Washington, DC: American Association for the Advancement of Science.

Furnham, A., & Gunter, B. (1987). Young people's political knowledge. *Educational Studies, 13,* 91–104.

Gallagher, J. J. (1991). Prospective and practicing secondary school science teachers' knowledge and beliefs about the philosophy of science. *Science Education, 75*(1), 121–133.

Gallagher, J., & Dawson, G. (Eds.). (1984). *Science education and cultural environments in the Americas.* Washington, DC: National Science Teachers Association.

Gauld, C. (1982). The scientific attitude and science education: A critical reappraisal. *Science Education, 66*(1), 109–121.

Gay, J., & Cole, M. (1967). *The new mathematics and an old culture. A study of learning among the Kpelle of Liberia.* New York: Holt, Rinehart & Winston.

George, J. M., & Glasgow, J. (1987, December). *Street science in two Carribean territories: Implications for teaching towards common science syllabi.* Paper presented at the Second Regional Consultation on Science Education in Latin America and the Caribbean, Kingston, Jamaica.

Gibbs, J., Arnold, L., & Burkhart, J. (1984). Sex differences in the expression of moral judgement. *Child Development,* (55), 1040–1043.

Gilligan, C. (1983). *In a different voice.* Cambridge, MA: Harvard University Press.

Gilliom, M. E., Helgeson, S. L., & Zuga, K. F. (Eds.). (1991). *Theory into Practice* [special issue on STS], *30*(4).

Gilliom, M. E., Helgeson, S. L., & Zuga, K. F. (Eds.). (1992). *Theory into Practice* [special issue on STS], *31*(1).

Glaser, B. G. (1969). The constant comparative method of qualitative analysis. In G. J. McCall & J. L. Simmons (Eds.), *Issues in participant observation: A text and reader* (pp. 216–227). Reading, MA: Addison-Wesley.

Glasser, R. (1991). The maturing of the relationship between the science of learning and cognition and educational practice. *Learning and Instruction, 1,* 129–144.

Globe and Mail [editorial]. (1983, March 2). Political science. *Globe and Mail,* p. 6.

Grant, M. (1987). Changing the polarity. *International Journal of Science Education, 9*(3), 335–342.

Grosvenor, G. (1988). Pandas in Panama? [President's page]. *National Geographic, 174*(5).

Hager, W. R., & Von Braun, M. (1989). Motivating minority and disadvantaged students in science and mathematics: Idaho summer camp. In *Pre-college education of minorities in science and engineering: Conference proceedings.* Newark, NJ: New Jersey Institute of Technology.

Hall, W. (1973). *Patterns: Teacher's handbook.* London: Longman.

Haraway, D. (1978). Animal sociology and the natural economy of the body politic (Pts. 1 and 2). *Signs: Journal of Women in Culture and Society, 4*(1), 21–60.

Haraway, D. (1990). *Primate visions.* New York: Reidel.

Harding, J. (1983). *Switched-off: The science education of girls.* London: Longman (for the Schools Council).

Harding, J. M., Kelly, P. J., & Nicodemus, R. B. (1976). The study of curriculum change. *Studies in Science Education, 3,* 1–32.

Harding, S. (1991). *Whose science? Whose knowledge?: Thinking from women's lives.* Ithaca, NY: Cornell University Press.

Harré, R. (1983). *Personal being.* Oxford: Basil Blackwell.

Harrison, G. B. (1980). The role of technology in science education. In C. McFadden (Ed.), *World trends in science education* (pp. 18–26). Halifax, Nova Scotia, Canada: Atlantic Institute of Education.

Hart, E. P. (1989). Toward renewal of science education: A case study of curriculum policy development. *Science Education, 73*(5), 607–634.

Hart, E. P., & Robottom, I. M. (1990). The science-technology-movement in science education: A critique of the reform process. *Journal of Research in Science Teaching, 27*(6), 575–588.

Hartsock, N. (1983). The feminist standpoint. In S. Harding & M. Hintikka (Eds.), *Discovering reality* (pp. 283–310). New York: Reidel.

Haste, H. (1993). Moral creativity and education for citizenship. *Creativity Research Journal, 6,* (182), 153–169.

Head, J. (1985). *The personal response to science.* Cambridge: Cambridge University Press.

Head, J., & Ramsden, J. (1990). Gender, psychological type and science. *International Journal of Science Education, 12*(1), 115–121.

Hines, J., Hungerford, H., & Tomera, A. (1987). Analysis and synthesis of research on responsible environmental behaviour. *Journal of Environmental Education, 18*(1), 1–8.

Hodge, B., & Tripp, D. (1986). *Children and television.* Cambridge: Polity Press.

Hofstein, A., Aikenhead, G., & Riquarts, K. (1988). Discussions over STS at the fourth IOSTE symposium. *International Journal of Science Education, 10*(4), 357–366.

Holman, J. (1986). *Science and technology in society: General guide for teachers.* Hatfield, Hertfordshire, UK: Association for Science Education.

Holton, G. (1978). *The scientific imagination: Case studies.* Cambridge: Cambridge University Press.

Holton, G., Rutherford, J., & Watson, F. (1970). *The Project Physics Course.* New York: Holt, Rinehart & Winston.

Horton, R. (1970). African traditional thought and western science. In B. Wilson (Ed.), *Rationality* (pp. 188–212). Oxford: Basil Blackwell.

Horton, R. (1971). African traditional thought and western science. In M. F. E. Young (Ed.), *Knowledge and control: New directions for sociology of education* (pp. 208–266). London: Collier-Macmillan.

Hueftle, J. S., Rakow, S. J., & Welch, W. W. (1983). *Images of science.* Minneapolis, MN: University of Minnesota, Minnesota Research and Evaluation Center.

Hunt, J. A. (1988). SATIS approaches to STS. *International Journal of Science Education, 10*(4), 409–420.

Hunt, J. A., & Lenton, G. (1992). SATIS 16–19 on trial in the U.K. In R. E. Yager (Ed.), *Status of STS: Reform efforts around the world* (pp. 19–25). 1992 ICASE Yearbook. Knapp Hill, South Harting, Petersfield UK: ICASE.

Hurd, P. (1971). *New directions in teaching secondary school science.* Chicago: Rand McNally.

Hurd, P. (1986). Perspectives for the reform of science education. *Phi Delta Kappan, 67*(5), 353–358.

Hurd, P. (1989). Science education and the nation's economy. In A. B. Champagne, B. E. Lovitts & B. J. Calinger (Eds.), *Scientific literacy* (pp. 15–40). Washington, DC: American Association for the Advancement of Science.

Iozzi, L. (1979). *Moral judgment, verbal ability, logical reasoning ability, and environment issues.* Unpublished doctoral dissertation, Rutgers University, New Brunswick, New Jersey.

Iozzi, L. (1984). *Summary of research in environmental education.* Monographs in Environmental Education and Environmental Studies. ERIC Clearing-house of Science Mathematics and Environmental Education. Columbus, OH: Ohio State University.

Jacobson, J. F. (1983). When scientists testify for hire. *Science 83, 4*(6), 18.

Jegede, O. J. (1988). The development of the science, technology and society curricula in Nigeria. *International Journal of Science Education, 10*(4), 399–408.

Jegede, O. J., & Fraser, B. J. (1989). Influence of socio-cultural factors on secondary school students attitude towards science. *Research in Science Education, 19,* 155–163.

Jegede, O. J., & Okebukola, P. A. O. (1988). The educology of socio-cultural factors in science classrooms. *International Journal of Educology, 2*(2), 93–107.

Jegede, O. J., & Okebukola, P. A. O. (1989). Some socio-cultural factors militating against drift towards science and technology in secondary schools. *Research in Science and Technological Education,* 7(2), 141–151.

Jegede, O. J., & Okebukola, P. A. O. (1990). The relationship between African traditional cosmology and student's acquisition of a science process skill. *International Journal of Science Education, 12*(1), 37–47.

Jegede, O. J., & Okebukola, P. A. O. (1991). The effect of instruction on socio-cultural beliefs hindering the learning of science. *Journal of Research in Science Teaching, 28*(3), 275–285.

Jegede, O. J., & Okebukola, P. A. O. (1993). Measuring the effects of socio-cultural factors in non-western science classrooms. *The Hong Kong Journal of Educational Research, 8,* 40–47.

Jenkins, E. (1990, April). *Domestic energy and the elderly: The understanding of energy.* Paper presented at the Public Understanding of Science conference, London Science Museum.

Jesuitas, P. P. (1988). Science and technology for living. In D. Layton (Ed.), *Innovations in science and technology education* (Vol. 2) (pp. 167–179). Paris: UNESCO.

Johnson, S., & Murphy, P. (1986). The underachievement of girls in physics: Towards explanations. *European Journal of Science Education, 8*(4), 401–411.

Jörg, T., & Wubbels, Th. (1987). Physics a problem for girls, or girls a problem for physics? *International Journal of Science Education, 9*(3), 297–307.

Kahle, J. B. (1987). SCORES: A project for change? *International Journal of Science Education, 9*(3), 325–333.

Kahle, J. B. (1988). Gender and science education (Pt. 2). In P. J. Fensham (Ed.), *Development and dilemmas in science education* (pp. 249–265). New York: Falmer Press.

Keeves, L. P. (1990, June). *Science education: Towards the future.* Paper presented at the First Major STS Conference in South America, Ministry of Education, Brasilia, Brazil.

Keller, E. F. (1977). The anomaly of a woman in physics. In S. Ruddick & P. Daniels (Eds.), *Working it out: 23 women, writers, scientists and scholars talk about their lives* (pp. 77–91). New York: Pantheon Books.

Keller, E. F. (1985). *Reflections on gender and science.* New Haven, CT: Yale University Press.

Kelly, A. (1981). *The missing half: Girls and science.* Manchester: Manchester University Press.

Kelly, G. J., Carlsen, W. S., & Cunningham, C. M. (1993). Science education in sociocultural context: Perspectives from the sociology of science. *Science Education, 77*(2), 207–220.

Kelly, P. (1991). Perceptions and performance: An impact assessment of CEPUP in schools. Lawrence Hall of Science, Berkeley, CA: Chemical Education for Public Understanding Program (CEPUP).

Klopfer, L. E. (1966). *History of science cases.* Chicago: Science Research Associates.

Klopfer, L. E., & Cooley, W. W. (1963). "The History of Science Cases" for high school in the development of student understanding of science and scientists. *Journal of Research in Science Teaching, 1*(1), 33–47.

Knamiller, G. (1989). Linking school science and technology with school science in Malawi. *Science Education Newsletter,* (84), 1–3.

Kohlberg, L. (1984). *Child psychology and childhood education.* New York: Longman.

Kohlberg, L., & Candee, D. (1984). The relationship of moral judgement to moral action. In L. Kohlberg (Ed.), *Essays on moral development* (Vol. 2). San Francisco: Harper & Row.

Kortland, K. (1992). Environmental education: Sustainable development and decision making. In R. E. Yager (Ed.), *The status of STS: Reform efforts around the world* (pp. 32–39). ICASE 1992 Yearbook. Knapp Hill, South Harting, Petersfield, UK: International Council of Associations for Science Education.

Kranzberg, M. (1991). Science-technology-society: It's as simple as XYZ! *Theory in Practice, 30*(4), 234–241.

Kuhn, T. (1970). *The structure of scientific revolutions* (2nd ed.). Chicago: University of Chicago Press.

Lakatos, I. (1970). Falsification and the methodology of scientific research programmes. In I. Lakatos & A. Musgrave (Eds.), *Criticism and the growth of knowledge* (pp. 91–196). Cambridge: Cambridge University Press.

Lambert, H., & Rose, H. (1990, April). *Disembodied knowledge? Making sense of medical knowledge.* Paper presented at the Public Understanding of Science conference, London Science Museum.

Larochelle, M., & Désautels, J. (1991). "Of course, it's obvious": Adolescents' ideas of scientific knowledge. *International Journal of Science Education, 13*(4), 373–389.

Latour, B., & Woolgar, S. (1979). *Laboratory life: The social construction of scientific facts*. London: Sage.

Layton, D. (1986). Empowerment of people: The educational challenge of science for specific social purposes (SSSP). *Bulletin of Science, Technology & Society, 6*(2/3), 210–218.

Layton, D. (1988). Revaluing the T in STS. *International Journal of Science Education, 10*(4), 367–378.

Layton, D. (1991). Science education and praxis: The relationship of school science to practical action. *Studies in Science Education, 19,* 43–79.

Layton, D. (Ed.) (1993a). *Innovations in science and technology education* (Vol. 5). Paris: UNESCO.

Layton, D. (1993b). *Technology's challenge to science education.* Buckingham, UK: Open University Press.

Layton, D., Davey, A., & Jenkins, E. (1986). Science for specific social purpose (SSSP): Perspectives on adult scientific literacy. *Studies in Science Education, 13,* 27–52.

Lederman, N. G. (1992). Students' and teachers' conceptions of the nature of science: A review of the research. *Journal of Research in Science Teaching, 29*(4), 331–359.

Leopold, A. (1949). *A sand county almanac.* Oxford: Oxford University Press.

Lepkrowski, W. (1989). Scholars ponder the goals of science, technology and society programs. *Chemical & Engineering News, 67*(12), 14–15.

Lewis, J. (1978). Teaching the relevance of science for society. In J. Lewis & P. J. Kelly (Eds.), *Science and technology education and future human needs* (pp. 19–24). London: Pergamon.

Lewis, J. (Proj. Dir.). (1981). *Science in society.* London: Heinemann Educational Books.

Lewis, J. (1985). Science and society. In G. B. Harrison (Ed.), *World trends in science and technology education* (pp. 29–30). Nottingham, UK: Trent Polytechnic.

Lijnse, P. (1990). Energy between the life-world of pupils and the world of physics. *Science Education, 74*(5), 571–583.

Lijnse, P. L., Eijkelhof, H. M. C., Klaassen, C. W. J. M., & Scholte, R. L. J. (1990). Pupils' and mass-media ideas about radioactivity. *International Journal of Science Education, 12*(1), 67–78.

Lloyd, A. (1990, August 28). A discovery that came too late. *Independent,* p. 19.

Lloyd, A., & Hall, C. (1990, August 28). Breakthrough in gene defects gives hope to children. *Independent,* p. 3.

Longino, H. (1990). *Science as social knowledge: Values and objectivity in scientific inquiry.* Princeton, NJ: Princeton University Press.

Lovelock, J. (1979). *Gaia: A new look at life on earth.* Oxford: Oxford University Press.

Lucas, A. M. (1983). Scientific literacy and informal education. *Studies in Science Education, 10,* 1–36.

Lucas, A. M. (1985). Who needs to know what about weather? In J. M. Walker

(Ed.), *Weather education: Proceedings of the first international conference on school and popular meteorological education* (pp. 233–247). Bracknell, UK: Royal Meteorological Society.

Lucas, A. M. (1987a). Public knowledge of biology. *Journal of Biological Education, 21,* 41–45.

Lucas, A. M. (1987b). Public knowledge of radiation. *Biologist, 34,* 125–129.

Lucas, A. M. (1987c). Interactions between formal and informal sources of learning science. In D. Evered and M. O'Connor (Eds.), *Communicating science to the public* (pp. 64–80). Chichester, UK: John Wiley & Sons, Ciba Foundation Conference.

Lucas, A. M. (1988). Public knowledge of elementary physics. *Physics Education, 23,* 10–16.

Lucas, A. M. (1989). Public understanding of science and the "official" English model of science. In D. E. Herget (Ed.), *The history and philosophy of science in science teaching* (pp. 230–241). Tallahasee, FL: Florida State University, Science Education and Department of Philosophy.

Lucas, A. M. (1991). 'Info-tainment' and informal sources for learning science. *International Journal of Science Education, 13*(4), 495–504.

Lucas, A. M., McManus, P. M., & Thomas, G. (1986). Investigating learning from informal sources: Listening to conversations and observing play in science museums. *European Journal of Science Education, 8*(4), 341–352.

Macgill, S. (1987). *The politics of anxiety.* London: Pion.

MacIntyre, A. (1985). *After virtue: A study in moral theory* (2nd ed.). London: Duckworth.

Macoby, E., & Jacklin, C. (1975). *The psychology of sex differences.* London: John Wiley & Sons.

Maddock, M. N. (1981). Science education: An anthropological view-point. *Studies in Science Education, 8,* 1–26.

Majumdar, S. K., Rosenfeld, L. M., Rubba, P. A., Miller, E. W., & Schmalz, R. F. (Eds.). (1991). *Science education in the United States: Issues, crises and priorities.* Easton, PA: Pennsylvania Academy of Science.

Malcolm, C. K., Cole, J., Hogendoorn, B., O'Keefe, D., & Reid, I. (1987). *The science framework P–10.* Melbourne, Victoria: Ministry of Education, Schools Division.

Malinowski, B. (1948). *Magic, science and religion.* Boston: Beacon Press.

Marks Greenfield, P. (1984). *Mind and media.* London: Fontana.

Matthews, M. R. (1990). History, philosophy and science teaching: A rapprochement. *Studies in Science Education, 18,* 25–51.

Mbiti, J. S. (1969). *African religion and philosophy.* London: Heinemann.

McCammon, S., Golden, J., & Wuensch, K. L. (1988). Predicting course performance in freshman and sophomore physics courses: Women are more predictable than men. *Journal of Research in Science Teaching, 25*(6), 501–510.

McConnell, M. C. (1982). Teaching about science, technology and society at the secondary school level in the United States: An educational dilemma for the 1980s. *Studies in Science Education, 9,* 1–32.

McCormack, M. N., & Yager, R. E. (1989). A new taxonomy of science education. *Science Teacher, 56*(2), 47–48.

McGinn, R. E. (1991). *Science, technology, and society.* Englewood Cliffs, NJ: Prentice Hall.

McManus, P. M. (1988). Do you get my meaning? Perception, ambiguity and the museum visitor. *ILVS Review, 1,* 62–75.

McPeck, J. E. (1981). *Critical thinking and education.* Oxford: Martin Robertson Press.

McQuail, D. (1984). *Communication.* Harlow, UK: Longman.

Meadows, D. H., Meadows, D. L., Randers, J., & Behrens, W. W. (1972). *The limits to growth.* Washington, DC: Potomac Associates.

Mendelsohn, E. (1975a). *Contemporary problems of science.* Hyattsville, MD: Current Information Associates.

Mendelsohn, E. (1975b). *Science in the 19th century.* Hyattsville, MD: Current Information Associates.

Mendelsohn, E. (1975c). *The emergence of modern science.* Hyattsville, MD: Current Information Associates.

Mendelsohn, E. (1976). Values and science: A critical reassessment. *Science Teacher, 43*(1), 20–23.

Merton, R. (1942). Science and technology in a democratic order. *Journal of Legal and Political Sociology, 1,* 115–126.

Mesaros, R. A. (1988). The effect of teaching strategies on the acquisition and retention of knowledge. (Doctoral dissertation, University of Pennsylvania, 1987). *Dissertation Abstracts International, 49*(2), 229A.

Meyers, H. M. (1992). STS and science concepts. In R. E. Yager (Ed.), *The status of STS: Reform efforts around the world* (pp. 76–80). ICASE 1992 Yearbook. Knapp Hill, South Harting, Petersfield, UK: International Council of Associations for Science Education.

Michael, M. (1992). Lay discourses of science: Science-in-general, science-in-particular and self. *Science, Technology & Human Values, 17*(3), 313–333.

Middleton, J., Ziderman, A., & Van Adams, A. (1991). *Vocational and technical education and training.* A World Bank Policy Paper. Washington, DC: World Bank.

Middleton, W. E. K. (1963). *The scientific revolution.* Toronto: Canadian Broadcasting Corporation.

Millar, R., & Wynne, B. (1988). Public understanding of science: From contents to processes. *International Journal of Science Education, 10*(4), 388–398.

Miller, J. D. (1983). Scientific literacy: A conceptual and empirical review. *Daedalus, 96*(1), 29–48.

Miller, J. D. (1989, October). *Who is scientifically literate?* Paper presented to the AAAS Science Education Forum, American Association for the Advancement of Science, Washington, DC.

Mitschke, V. J. (1993). *Teacher practical knowledge: Teachers' reasons for selecting instructional methods.* Unpublished master's thesis, University of Saskatchewan, Saskatoon, Canada.

Moore, J. H. (1991). Science education, the national economy, and international competitiveness. In S. K. Majumdar, L. M. Rosenfeld, P. A. Rubba, E. W. Miller, & R. F. Schmalz (Eds.), *Science education in the United States* (pp. 528–531). Easton, PA: Pennsylvania Academy of Science.

Moscovici, S. (1976). *Social influence and social change.* London: Academic Press.

Mukund, K. (1988, October 15). The Hoshangabad science teaching programme. *Economic and Political Weekly,* pp. 2147–2150.

Mulkay, M. (1975). Norms and ideology in science. *Social Science Information, 15,* 637–656.

Murphy, P. (1990, January). Gender gap in the national curriculum. *Physics World,* pp. 11–12.

National Science Board. (1988). *Science indicators.* Washington, DC: U.S. Government Printing Office.

National Science Foundation (NSF). (1983). *Educating Americans for the 21st century.* Washington, DC: National Science Board.

National Science Foundation (NSF). (1990). NSF to support statewide reforms in science, math, and engineering education. *NSF Directions, 3*(3), 1.

National Science Teachers Association (NSTA). (1982). Science-technology-society: Science education for the 1980s (NSTA Position Statement). Washington, DC: Author.

National Science Teachers Association (NSTA). (1990). Position paper on STS. *Bulletin of Science, Technology & Society, 10*(5, 6), 249–250.

National Science Teachers Association (NSTA). (1991, October/November). NSTA position statement: Laboratory science. *NSTA Reports.*

Nganunu, M. (1988). An attempt to write a science curriculum with social relevance for Botswana. *International Journal of Science Education, 10*(4), 441–448.

Northern Examination Board. (1988). *Science, technology and society: Syllabus for the 1988 GCSE examination.* Manchester, UK: Associated Lancashire Schools Examining Board.

Nuffield Modular Sciences (1992–1993). *Pathways through science.* (A series of 13 modules). London: Longman.

Nuffield Physics. (1967). *Teachers' guide 4.* London: Longman.

Nuffield Secondary Science. (1971). *Teachers' guide.* London: Longman.

Odhiambo, T. R. (1972). Understanding of science: The impact of the African view of nature. In P. G. S. Gilbert & M. N. Lovegrove (Eds.), *Science education in Africa.* London: Heinemann.

Ogawa, M. (1989). Family-based STS education: A new approach. *Bulletin of Science, Technology & Society, 9*(4), 239–244.

Ogbu, J. U. (1992). Understanding cultural diversity and learning. *Educational Researcher, 21*(8), 5–14, 24.

Ogden, W. R. (1975). Secondary school chemistry teaching, 1918–1972: Objectives as stated in periodical literature. *Journal of Research in Science Teaching, 12*(3), 235–246.

Ogunniyi, M. B. (1988a). Adapting western science to traditional African culture. *International Journal of Science Education, 10*(1), 1–9.

Ogunniyi, M. B. (1988b, September). *Sustaining students' interests in science and technology: The socio-cultural factors.* Lead paper presented at the 29th Annual Conference of the Science Teachers' Association of Nigeria, Ibadan.

Okebukola, P. A. O., & Jegede, O. J. (1990). Eco-cultural influences on students' concept attainment in science. *Journal of Research in Science Teaching, 27*(7), 661–669.

Olson, J. (Ed.) (1982). *Innovation in the science curriculum.* London: Croom Helm.
Oppenheimer, F. (1985). Teaching and learning [Special memorial issue]. *Exploratorium, 9,* 8–9.
Orpwood, G. (1985). Toward the renewal of Canadian science education: Vol. 1. Deliberative inquiry model. *Science Education, 69*(4), 477–489.
Oxford University Department of Educational Studies. (1989). *Enquiry into the attitudes of sixth-formers towards choice of science and technology courses in higher education.* Oxford: Oxford University, Department of Educational Studies.
Paden, R. (1992). Nature and morality. *Environmental Ethics, 14*(3), 239–251.
Passmore, J. (1974). *Man's responsibility for nature.* London: Duckworth.
Peaker, C. (1967). *Children and their primary school.* Plowden Committee Report. London: HMSO.
Pedersen, J. E. (1992). The jurisprudential model of study for STS issues. In R. E. Yager (Ed.), *The status of STS: Reform efforts around the world* (pp. 26–31). ICASE 1992 Yearbook. Knapp Hill, South Harting, Petersfield, UK: International Council of Associations for Science Education.
Pell, A. (1985). Enjoyment and attainment in secondary school physics. *British Education Research Journal, 11*(2), 123–132.
Penick, J. E. (1992). STS instruction enhances student creativity. In R. E. Yager (Ed.), *The status of STS: Reform efforts around the world* (pp. 81–86). ICASE 1992 Yearbook. Knapp Hill, South Harting, Petersfield, UK: International Council of Associations for Science Education.
Piel, E. P. (1981). Interaction of science, technology, and society in secondary schools. In N. Harms & R. Yager (Eds.), *What research says to the science teacher* (Vol. 3, pp. 94–112). Washington, DC: National Science Teachers Association.
Piel, J. E. (1993). Decision making: A goal of STS. In R. E. Yager (Ed.), *The science, technology, society movement* (pp. 147–152). Washington, DC: National Science Teachers Association.
Pinch, T. J. (1992). Opening black boxes: Science, technology and society. *Social Studies of Science, 22*(3), 487–510.
PLON (1984). *Ioniserende Straling.* Utrecht: Rijksuniversiteit Utrecht, Vakgroep Natuurkunde-Didactiek. English version (1988): *Ionizing Radiation.* Melbourne: Monash University, Faculty of Education.
Popper, K. R. (1959). *The logic of scientific discovery.* New York: Harper & Row.
Popper, K. R. (1963). *Conjectures and refutations: The growth of scientific knowledge.* London: Routledge & Kegan Paul.
Posner, G. J. (1992). *Analyzing the curriculum.* New York: McGraw-Hill.
Price, D. (1963). *Little science, big science.* New York: Columbia University Press.
Rampal, A. (1991). Deliverance from the 'delivery' metaphor: Curriculum innovation in India. *Journal of Education for Teaching, 17*(3), 237–244.
Rampal, A. (1992a). A possible 'orality' for science? *Interchange, 23*(3), 227–244.
Rampal, A. (1992b). School science in search of a democratic order? *Social Scientist, 20*(7/8), 50–74.
Rampal, A. (1992c). The image of science and scientists: A study of school teachers' views. I. Characteristics of scientists. *Science Education, 76*(4), 415–436.

Ramsay, J. (1993). The science education reform movement: Implications for social responsibility. *Science Education, 72*(2), 235–258.

Ramsden, J. M. (1992). If it's enjoyable, is it science? *Science Science Review, 73*(265), 65–71.

Ravetz, J. R. (1971). *Scientific knowledge and its social problems*. New York: Oxford University Press.

Rawls, J. (1971). *A theory of justice*. Cambridge, MA: Harvard University Press.

Razali, S. (1986). *Comparison of perceptions of the importance of high school chemistry among various instructors and students in the United States and Malaysia*. Unpublished doctoral dissertation, University of Iowa, Iowa City.

Riesch, W., & Westphal, W. (1975). Modellhafte schülervorstellungen zur ausbreitung radioaktiver strahlung. *Der Physikunterricht, 9*(4), 75–85.

Roberts, D. A. (1980). Theory, curriculum development, and the unique events of practice. In H. Munby, G. Orpwood, & T. Russell (Eds.), *Seeing curriculum in a new light: Essays from science education* (pp. 65–87). Toronto: OISE Press.

Roberts, D. A. (1983). *Scientific literacy*. Ottawa: Science Council of Canada.

Roberts, D. A. (1988). What counts as science education? In P. J. Fensham (Ed.), *Development and dilemmas in science education* (pp. 27–54). New York: Falmer Press.

Roberts, H. (Ed.). (1981). *Feminist methodology*. London: Routledge & Kegan Paul.

Roeder, J. L. (1993). *Teachers Clearinghouse Newsletter.* (Available from 1 West 88 St, New York, 10024.)

Rose, H. (1979). Hyper-reflexivity: A new danger for the countermovements. In H. Nowotny & H. Rose (Eds.), *Countermovements in the sciences: Sociology of science yearbook* (pp. 277–289). Dordrecht: Reidel.

Rose, H. (in press). *Love, power and knowledge: Feminist science criticism and feminist science theory*. Cambridge: Polity Press.

Rose, H., & Rose, S. (1979). The myth of the neutrality of science. In R. Arditti, P. Brennan, & S. Cavrak (Eds.), *Science and liberation* (pp. 17–32). Boston: South End Press.

Rosenberg, A. (1985). *The structure of biological science*. Cambridge: Cambridge University Press.

Rosenthal, D. B. (1989). Two approaches to STS education. *Science Education, 73*(5), 581–589.

Roszak, T. (1972). *Where the wasteland ends*. London: Faber & Faber.

Roth, W-M., & Roychoudhury, A. (1993). The nature of scientific knowledge, knowing, and learning: The perspectives of four physics students. *International Journal of Science Education, 15* (1), 27–44.

Roy, R. (1988). *S-STS project final report* (NSF #SPE-8317984). Washington, DC: National Science Foundation.

Roy, R., & Barchi, B. (1990). *National STS network final report* (NSF # TPE-8751239). Washington, DC: National Science Foundation.

Rubba, P. A. (1987a). Perspectives on science-technology-society instruction. *School Science and Mathematics, 87*(3), 181–185.

Rubba, P. A. (1987b). The current state of research in precollege STS education: A position paper. *Bulletin of Science, Technology & Society,* 7(1/2), 248–252.

Rubba, P. A. (1991). Integrating STS into school science and teacher education: Beyond awareness. *Theory into Practice, 30*(4), 303–308.

Rubba, P. A., McGuyer, M., & Wahlund, T. M. (1991). The effect of infusing STS vignettes into the genetics unit of biology on learner outcomes in STS and genetics. *Journal of Research in Science Teaching, 28*(6), 537–552.

Rubba, P. A., & Wiesenmayer, R. L. (1985a). A goal structure for precollege STS education: A proposal based upon recent literature in environmental education. *Bulletin of Science, Technology & Society, 5*(6), 573–580.

Rubba, P. A., & Wiesenmayer, R. L. (1985b). The implications of recent research in environmental education for Science-Technology-Society. *Bulletin of Science, Technology & Society, 5*(6), 573–580.

Runciman, W. G. (1989). *A treatise on social theory.* Cambridge: Cambridge University Press.

Rutherford, F. J. (1988). STS . . . Here today and . . . ? *Bulletin of Science, Technology & Society, 8*(2), 126–127.

Ryan, A. G., & Aikenhead, G. S. (1992). Students' preconceptions about the epistemology of science. *Science Education, 76*(6), 559–580.

Salomon, J-J. (1973). *Science and politics.* London: Macmillan.

Sayre, A. (1975). *Rosalind Franklin and DNA: A vivid view of what it is like to be a gifted woman in an especially male profession.* New York: W. W. Norton.

Schools Council Integrated Science Project. (1973). *Patterns.* London: Longman.

Schutz, A., & Luckmann, T. (1973). *Structures of the life world.* London: Heinemann.

Schwab, J. J. (1974). Decision and choice: The coming duty of science teaching. *Journal of Research in Science Teaching, 11*(4), 309–317.

Schwab, J. J. (1978). *Science, curriculum, and liberal education.* Chicago: University of Chicago Press.

Science Council of Canada. (1984). *Science for every student: Educating Canadians for tomorrow's world* (Report No. 36). Ottawa, Ontario: Science Council of Canada.

Science Field of Study Committee. (1989). *Chemistry in the science field.* Melbourne, Victoria, Australia: Victorian Curriculum and Assessment Board.

Science Through STS Project. (1985). Task force establishes materials review process. *S-STS Reporter, 1*(3), 5–6.

Science Through STS Project. (1986). A one-year course in STS. *S-STS Reporter, 2*(4), 7–21.

Secretary's Commission on Achieving Necessary Skills (SCANS). (1991). *What work requires of schools.* Washington, DC: U.S. Government Printing Office.

SEPUP Annual Report (1992). *Science education for public understanding program.* Berkeley, CA: University of California at Berkeley.

Shapiro, B. L. (1989). What children bring to light: Giving high status to learners' views and actions in science. *Science Education, 73*(6), 711–733.

Shipman, M. D. (1972). Contrasting views of a curriculum project. *Journal of Curriculum Studies, 4*(2), 145–153.

Shymansky, J. A., Hedges, L. V., & Woodworth, G. (1990). A reassessment of inquiry-bases science curricula of the 60's on student performance. *Journal of Research in Science Teaching, 27*(2), 127–144.

Silverstone, R. (1985). *Framing science: The making of a BBC documentary.* London: British Film Institute.

Sjøberg, S., & Imsen, G. (1988). Gender and science education. I. In P. J. Fensham (Ed.), *Development and dilemmas in science education* (pp. 218–248). New York: Falmer Press.

Skilbeck, M. (1984). *School-based curriculum development.* London: Harper.

Smail, B. (1984). *Girl-friendly science: Avoiding sex bias in the curriculum.* London: Longman (for the Schools Council).

Snow, R. E. (1987). Core concepts for science and technology literacy. *Bulletin of Science, Technology & Society,* 7(5/6), 720–729.

Solomon, J. (1981). Science and society studies in the curriculum. *School Science Review,* (82), 213–220.

Solomon, J. (1983). *Science in a social context* (SISCON)*-in-schools.* Oxford: Basil Blackwell.

Solomon, J. (1984). Prompts, cues and discrimination: The utilization of two separate knowledge systems. *European Journal of Science Education, 6*(3), 277–284.

Solomon, J. (1985). Learning and evaluation: A study of school children's views on the social uses of energy. *Social Studies of Science, 15*(2), 343–371.

Solomon, J. (1987). Social influences on the construction of pupil's understanding of science. *Studies in Science Education, 14,* 63–82.

Solomon, J. (1988a). Science technology and society courses: Tools for thinking about social issues. *International Journal of Science Education, 10*(4), 379–387.

Solomon, J. (1988b). The dilemma of science, technology and society education. In P. J. Fensham (Ed.), *Development and dilemmas in science education* (pp. 266–281). New York: Falmer Press.

Solomon, J. (1989). The social construction of school science. In R. Millar (Ed.), *Doing science: Images of science in science education* (pp. 126–136). New York: Falmer Press.

Solomon, J. (1991). *Exploring the nature of science.* London: Blackie & Son.

Solomon, J. (1992a). *Getting to know about energy.* London: Falmer Press.

Solomon, J. (1992b). *How does society decide?* Hatfield, Hertfordshire, UK: Association for Science Education.

Solomon, J. (1992c). The classroom discussion of science-based social issues presented on television: Knowledge, attitudes and values. *International Journal of Science Education, 14*(4), 431–444.

Solomon, J. (1992d). *What is science?* Hatfield, Hertfordshire, UK: Association for Science Education.

Solomon, J. (1992e). *What is technology?* Hatfield, Hertfordshire, UK: Association for Science Education.

Solomon, J. (1993). *Teaching science, technology and society.* Buckingham, UK: Open University Press.

Solomon, J., Duveen, J., Scot, L., & McCarthy, S. (1992). Teaching about the nature of science through history: Action research in the classroom. *Journal of Research in Science Teaching, 29*(4), 409–421.

Solomon, J., & Harrison, K. (1991). Talking about science based issues: Do boys and girls differ? *British Educational Research Journal, 17* (3), 283–294.

Spector, R. S., & Gibson, C. W. (1991). A qualitative study of middle school students' perception of factors facilitating learning of science: Grounded theory and existing practice. *Journal of Research in Science Teaching, 28*(2), 111–121.

Spiegel-Rösing, I. (1977). The study of science, technology and society (SSTS): Recent trends and future challenges. In I. Spiegel-Rösing & D. Price (Eds.), *Science, technology and society: A cross-disciplinary perspective* (pp. 7–42). London: Sage.

Stanley, J. C., & Stanley, B. S. (1986). High school biology, chemistry, or physics learned well in three weeks. *Journal of Research in Science Teaching, 23*(3), 237–250.

Stéhelin, L. (1976). Sciences, women and ideology. In H. Rose & S. Rose (Eds.), *The radicalisation of science* (pp. 76–89). London: Macmillan.

Strauss, A. L. (1987). *Qualitative analysis for social scientists.* Cambridge: Cambridge University Press.

Strike, K., & Posner, G. (1982). Conceptual change and science teaching. *European Journal of Science Education, 4*(3), 231–240.

Stuart, T. C. (1977). A comparison of high school and college chemistry courses in New Mexico. *Journal of Chemical Education, 54,* 373–374.

Susilo, H. (1976). *A comparison of perceptions of the importance of high school biology among various instructors and students in the United States and Indonesia.* Unpublished doctoral dissertation, University of Iowa, Iowa City.

Sutton, C. (1992). *Words, science and learning.* Buckingham, UK: Open University Press.

Tanaka, J., & Taigen, J. (1986, July/August). *Predictability of college chemistry grades based on high school variables.* Paper presented at the 9th Biennial Conference on Chemical Education, Montana State University, Bozeman.

Taylor, P. (1986). *Respect for nature.* Princeton, NJ: Princeton University Press.

Thatcher, M. (1971, November). *The essentials of a good education in a technological age.* Paper presented at the Seventeenth Annual Ford Lecture to Youth, Royal Festival Hall, London.

Thier, H. D. (1991). 4–2–1 CEPUP Fellows. Berkeley, CA: University of California at Berkeley.

Thier, H. D., & Hill, T. (1988). Chemical education in schools and the community: The CEPUP project. *International Journal of Science Education, 10*(4), 421–430.

Thomas, P. A. (1985). *Teaching rational argument skills in the context of science-related world problems.* Unpublished master's thesis, University of Saskatchewan, Saskatoon, Saskatchewan, Canada.

Thomson, T. E. (1975). *A survey of the status and needs opinion of science education in Oregon during the 1973–74 school year.* Unpublished doctoral dissertation, Oregon State University, Corvallis.

Tobias, S. (1990). *They're not dumb, they're different.* Tucson, AZ: Research Corporation.

Tulley, A. (1990). Seeing through the name. *International Journal of Museum Management and Curatorship, 9,* 53–62.

UNESCO. (1993). Project 2000+: An international project on scientific and technological literacy for all. Paris: UNESCO.

Urevbu, A. O. (1987a). School science in West Africa: An assessment of the pedagogical impact of third world investment. *International Journal of Science Education, 9*(1), 3–12.

Urevbu, A. O. (1987b). Cross-cultural teaching of the interaction of science: An African perspective. In I. Lowe (Ed.), *Teaching the interactions of science, technology and society* (pp. 283–293). Melbourne, Victoria: Longman Cheshire Pty.

UYSEG. (1991). *Salters' science.* York, UK: University of York Science Education Group (UYSEG).

Varella, G. F. (1992). Greater ability to apply concepts using an STS approach to teaching science. In R. E. Yager (Ed.), *The status of STS: Reform efforts around the world* (pp. 87–92). ICASE 1992 Yearbook. Knapp Hill, South Harting, Petersfield, UK: International Council of Associations for Science Education.

Visavateeranon, S., & Finley, F. N. (1993, April). *Effect of research experiences on teachers' perceptions of the nature of science.* Paper presented to the annual meeting of the National Association of Research in Science Teaching, Atlanta, GA.

Vygotsky, L. S. (1978). *Mind in society: The development of higher psychological processes.* Cambridge, MA: Harvard University Press.

Waks, L. J. (1987a). A technological literacy credo. *Bulletin of Science, Technology & Society,* 7(1/2), 357–366.

Waks, L. J. (1987b). Afterword: The STS prophets and their challenge to STS education. *Bulletin of Science, Technology & Society,* 7(5/6), 1001–1008.

Waks, L. J., & Prakash, M. S. (1985). STS education and its three step-sisters. *Bulletin of Science, Technology & Society, 5*(2), 105–116.

Walberg, H. J. (1991). Improving school science in advanced and developing countries. *Review of Educational Research, 61*(1), 25–69.

Wallsgrove, R. (1980). The masculine face of science. In Brighton Women and Science Group (Ed.), *Alice through the looking glass* (pp. 228–240). London: Virago.

Weart, S. R. (1988). *Nuclear fear: A history of images.* Cambridge, MA: Harvard University Press.

Weisstein, N. (1977). Adventures of a woman in science. In S. Ruddick & P. Daniels (Eds.), *Working it out: 23 women, writers, scientists and scholars talk about their lives* (pp. 241–250). New York: Pantheon Books.

Welch, W. W. (1969). Curriculum evaluation. *Review of Educational Research, 39*(3), 429–443.

Welch, W. W. (1973). Review of the research and evaluation program of Harvard Project Physics. *Journal of Research in Science Teaching, 10*(4), 365–378.

Welch, W. W., & Rothman, A. I. (1968). The success of recruited students in a new physics course. *Science Education, 52*(3), 270–273.

Whyte, J. (1986). *Girls into science and technology.* London: Routledge & Kegan Paul.

Wiesenmayer, R. L., Murrin, M., & Tomera, A. (1984). Environmental education research related to issue awareness. In L. A. Iozzi (Ed.), *Monographs in environmental education and environmental studies* (pp. 61–94). Columbus, OH: Ohio State University, ERIC Clearing-House of Science Mathematics and Environmental Education.

Wiesenmayer, R. L., & Rubba, P. A. (1990, April). *The effects of STS issue investigation and action instruction and traditional life science instruction on seventh grade stu-*

dents' citizenship behaviors. Paper presented at the annual meeting of the National Association for Research in Science Teaching, Atlanta, GA.

Wilsman, M. J. (1991). *Summative evaluation report. Phase 3. Secondary grades STS project.* Madison, WI: Wisconsin Public Telecommunications for Education.

Wingfield, L., & Haste, H., (1987). Connectedness and separateness: Cognitive style or moral orientation? *Journal of Moral Education, 16*(3), 214–225.

Wirth, A. G. (1991). Issues in the reorganization of work: Implications for education. *Theory in Practice, 30*(4), 280–287.

Wober, J. (1985). *The primacy of print over screen?* London: Independent Broadcasting Authority.

Woolf, V. (1952). *Three guineas.* London: Hogarth Press.

Wooster, D. (1977). *Nature's economy: A history of ecological ideas.* Cambridge: Cambridge University Press.

World Commission on Environment and Development. (1987). *Our common future: The Brundtland report.* Oxford: Oxford University Press.

Wynne, B. (1990, April 12). The blind and the blissful. *The Guardian,* p. 14.

Yager, R. E. (1983). In defense of societal issues as organizers for school science. *School Science and Mathematics, 83*(3), 651–653.

Yager, R. E. (1992a). Science-technology-society as reform. In R. E. Yager (Ed.), *The status of STS: Reform efforts around the world* (pp. 2–8). ICASE 1992 Yearbook. Knapp Hill, South Harting, Petersfield, UK: International Council of Associations for Science Education.

Yager, R. E. (Ed.) (1992b). *Status of STS: Reform efforts around the world.* 1992 ICASE Yearbook. Knapp Hill, South Harting, Petersfield, UK: International Council of Associations for Science Education.

Yager, R. E. (Ed.) (1992c). *The science, technology, society movement.* Washington, DC: National Science Teachers Association.

Yager, R. E., Blunck, S. M., Binadja, A., McComas, W. F., & Penick, J. E. (1988). *Assessing impact of STS instruction on 4–9 science in five domains* (ERIC Document Reproduction Service No. ED 292 641). Columbus, OH: Ohio State University, ERIC Clearing-House of Science Mathematics and Environmental Education.

Yager, R. E., & Krajcik, J. (1989). Success of students in a college physics course with and without experiencing a high school course. *Journal of Research in Science Teaching, 26*(7), 599–608.

Yager, R. E., & McCormack, A. J. (1989). Assessing teaching/learning successes in multiple domains of science and science education. *Science Education, 73*(1), 45–58.

Yager, R. E., Snider, B., & Krajcik, J. (1988). Relative success in college chemistry for students who experienced a high school course in chemistry and those who had not. *Journal of Research in Science Teaching, 25*(5), 387–396.

Young, M. (1971). An approach to the study of curricula as socially organized knowledge. In M. Young (Ed.), *Knowledge and control* (pp. 19–46). London: Collier-Macmillan.

Young, P., Ruck, C., & Crocker, B. (1991). Reading science. *Science Teacher, 58*(2), 46–49.

Zar'our, G. I. (1972). Superstitions among certain groups of Lebanese Arab students in Beirut. *Journal of Cross-Cultural Psychology, 3*(3), 273–282.

Ziman, J. (1980). *Teaching and learning about science and society.* Cambridge: Cambridge University Press.

Ziman, J. (1984). *An introduction to science studies: The philosophical and social aspects of science and technology.* Cambridge: Cambridge University Press.

Zoller, U., Ebenezer, J., Morely, K., Paras, S., Sandberg, V., West, C., Wolthers, T., & Tan, S. H. (1990). Goal attainment in science-technology-society (STS) education and reality: The case of British Columbia. *Science Education, 74*(1), 19–36.

Zoller, V., Donn, S., Wild, R., & Beckett, P. (1991). Students' versus their teachers' beliefs and positions on science-technology-society oriented issues. *International Journal of Science Education, 13*(1), 25–36.

About the Contributors

Glen S. Aikenhead is a Professor in the College of Education, University of Saskatchewan, Saskatoon, Canada. A humanistic perspective on science has guided his research into student assessment, classroom instruction, curriculum policy, and classroom materials. Dr. Aikenhead has published books, monographs, and articles in all four of the above areas, including *Science: A Way of Knowing; Science in Social Issues: Implications for Teaching; Teaching Science Through an STSE Approach;* and *Logical Reasoning in Science Technology.* He currently serves as chairperson of the International Organization for Science and Technology Education and as Special Section Editor for *Science Education.*

William S. Carlsen is an Associate Professor in the Department of Science Education, Cornell University. Professor Carlsen is known for his research interests in teacher education, educational sociolinguistics, and the effects of science teachers' subject matter knowledge on teaching.

Deborah J. Corrigan is Lecturer in Chemistry and Science Education, School of Graduate Studies, Faculty of Education, Monash University, Australia, and also has 10 years' experience as a chemistry and science education, industry and technology links with science education, and curriculum evaluation. She is a co-author of books on research findings in science education.

Christine M. Cunningham is a doctoral student at Cornell University. Her current research addresses science teachers' views on the sociological nature of scientific knowledge, and the role these views may play in the relationship between the teachers' subject matter knowledge and their teaching innovations.

Harrie M. C. Eijkelhof is a Senior Lecturer at the Centre of science and Mathematics Education, University of Utrecht, The Netherlands, Dr. Eijkelhof has published widely in the areas of teaching and learning with STS curricula and science textbook analysis. He is a co-author of the Dutch

Physics Curriculum Development Project's PLON materials and author of numerous articles.

Peter J. Fensham is Emeritus Professor of Science Education, School of Graduate Studies, Faculty of Education, Monash University, Australia, where he held a chair in Science Education from 1967 until his retirement in 1992. Professor Fensham's interests include science education (curriculum issues), environmental and technology education, and equity in education. His significant publications include *Science for All; Development and Dilemmas in Science Education;* and "Science and Technology" in the *Handbook of Research on Curriculum.*

Andrew Hunt is the Project Director of Nuffield Science in Practice, Nuffield Curriculum Projects Centre, London. He was a full-time science teacher for 20 years (1964–1984) and has contributed as a writer and editor to the revision of Nuffield Chemistry at the O and A levels. He is a founding member of the Science and Technology in Society Project (SATIS) project, a contributor to Salters Advanced Chemistry and author of a number of textbooks.

Olugbemiro J. Jegede is Head of the Research and Evaluation Unit, Distance Education Centre, University of Southern Queensland, Australia. Previously he was Head of Science Education and Assistant Dean of the Faculty of Education at Ahmadu Bello University, Nigeria, and foundation Professor and Dean of the Faculty of Education at the University of Abuja, Nigeria. Professor Jegede has researched and published extensively in STS, megacognition and science education, sociocultural aspects of science teaching and learning, and distance education.

Gregory J. Kelly is a doctoral student at Cornell University. His research interests include the study of conceptual change in science students, and consideration of the social context of scientific knowledge as related to science education.

David Layton is Emeritus Professor, Centre for Studies in Science and Mathematics Education, University of Leeds, England. He served as Director of the Centre from 1970 to 1982 and was a Professor in the university's Department of Science Education from 1973 to 1989. Professor Layton's principal research interest is the social history and politics of school science and technology education. The author of *Science for the People; Interpreters of Science; Technology's Challenge to Science Education* and co-author of *Inarticulate Science?,* Professor Layton is the editor of *Innovation in Science and Technology*

Education (Vols. 1–5, 1986–1993) for UNESCO and founder and past editor (1974–1982) of *Studies in Science Education.*

Arthur M. Lucas is the Principal of King's College London, England. Professor Lucas ia a former Head of the School of Education at both University of London's King's College and the University of South Australia. His research interests include history of science (especially botany), informal sources of learning science, and cognition and the learning of science. He most recently has coedited *Children's Informal Ideas in Science.*

Barbara W. Nagle is the Development Coordinator of the Science Education for Public Understanding Program (SEPUP), University of California, Berkeley. A Teacher Associate at the Lawrence Hall of Science, Dr. Nagle works with SEPUP through an agreement with the Oakland (California) Unified School District, where she taught high school chemistry for 6 years. Dr. Nagle is coordinating the development of SEPUP Course One.

Anita Rampal is working as a UGC Research Scientist at the Nehru Memorial Library, India. Dr. Rampal is currently engaged in research in the area of cognition and communication of science, with a special interest in the language of science. Involved in curriculum development and teacher training, she has taught graduate and undergraduate courses at the Department of Education, Jumia Millia University, New Delhi, and actively participated in and supported the Peoples' Science Movement in India.

Hilary Rose is Professor of Social Policy at the University of Bradford and was Director of the West Yorkshire Centre for Research on Women for 8 years. Professor Rose has researched and published extensively in STS, beginning with *Science and Society* in 1969; her forthcoming book is titled *Love, Power and Knowledge: Towards a Feminist Transformation of the Sciences.*

Joan Solomon is a Lecturer in Research in the Department of Educational Studies, Oxford University. She was the National Coordinator of the Science in a Social Context (SISCON) Project. Dr. Solomon has carried out substantial research on many aspects of Science education and is the author of *Teaching Children in the Laboratory, SISCON-In-Schools, Exploring the Nature of Science, Getting to Know About Energy, Teaching Science, Technology & Society,* and numerous articles about teaching science in its social context.

Herbert D. Thier is Associate Director of the Lawrence Hall of science at the University of California, Berkeley, and Director of its Science Education for Public Understanding Program (SEPUP). He is also the principal investigator and/or director of a number of externally funded national and international projects. His current work focuses on issue-oriented science that addresses risk understanding and evidence-based decision making in materials and strategies for the schools, community, and workplace. The senior author and one of the original developers of the Science Curriculum Improvement Study (SCIS) for the elementary schools, Dr. Thier is a Fellow of the American Association for the Advancement of Science.

William F. Williams is Professor of Science, Technology and Society and William Weiss Fellow at the Pennsylvania State University, and a Fellow of the Institute of Physics and the Institution of Electrical Engineers. A nuclear physicist and electronics engineer, Professor Williams is also a Life Fellow of the University of Leeds, England, where he served as Director of Combined Studies in Science until his retirement in 1983, and was the first Coordinator of the Science in a Social Context (SISCON) Project in the United Kingdom. Since 1984, Professor Williams has taken part in the Science through STS(S-STS) and National STS Network projects, and in the National Association of STS in the United States. He is co-editor of *Risk & Participation* and *The Bulletin of STS,* co-translator of *Physics in Society,* and author of *Are Science & Technology Neutral?* and "Broadening Courses," in *Learning Strategies in University Science.*

John Ziman is Emeritus Professor of Physics at the University of Bristol and a Fellow of the Royal Society of London. Professor Ziman's past positions and appointments of distinction include Professor of Theoretical Physics, Bristol University (1964–1982); Chairman, Council for Science & Society, London (1976–1990); Director, Science Policy Support Group, London (1986–1990); and Chairman, European Association for the Study of Science and Technology (1982–1986). Professor Ziman is the author of *Public Knowledge; The Force of Knowledge; Reliable Knowledge; Teaching and Learning about Science and Society; An Introduction to Science Studies;* and numerous articles on the social relations of science and technology.

Index